ADHESION 9

This volume is based on papers presented at the 22nd annual conference on Adhesion and Adhesives held at The City University, London

Previous conferences have been published under the titles of
Adhesion 1–8

ADHESION 9

Edited by

K. W. ALLEN

*Adhesion Science Group, Department of Chemistry,
The City University, London, UK*

ELSEVIER APPLIED SCIENCE PUBLISHERS
LONDON and NEW YORK

ELSEVIER APPLIED SCIENCE PUBLISHERS LTD
Ripple Road, Barking, Essex, England

Sole Distributor in the USA and Canada
ELSEVIER SCIENCE PUBLISHING CO., INC.
52 Vanderbilt Avenue, New York, NY 10017, USA

British Library Cataloguing in Publication Data:
Adhesion.—9
1. Adhesion
541.3′453   QD506

ISBN-13: 978-94-010-8688-2   e-ISBN-13: 978-94-009-4938-6
DOI: 10.1007/978-94-009-4938-6

WITH 46 TABLES AND 104 ILLUSTRATIONS

The selection and presentation of material and the opinions expressed in
this publication are the sole responsibility of the authors concerned.

Preface

The use of adhesives continues to increase both in importance and in areas of use, particularly in engineering construction. Alongside this is a steady growth in our fundamental understanding of the factors involved. Both of these topics are reflected in various meetings and nowhere more consistently than at the Annual Conference on Adhesion and Adhesives at The City University each Easter. 1984 was the twenty-second of these when, once again, people came from far and near to present accounts of their work and to discuss them in all their variety. This publication makes the papers available to the wider audience who were not able to attend in person.

In presenting this volume, may I express sincere gratitude, both personally and on behalf of the University, to all those who contributed in so many and various ways to the conference.

K. W. ALLEN

Contents

Chapter 1

TEXTILE-to-RUBBER ADHESION:

THEORY and PRACTICE

David B. WOOTTON

MILLIKEN INDUSTRIALS Ltd., BURY, LANCS.

1.   INTRODUCTION

The reinforcement of rubber with textiles plays a very important part in
the modern world; it is only necessary to consider the role of the tyre in
present day communication and transport to appreciate this. However, the full
performance of these composites can only be realised if the adhesion between
the various components is adequate.

In the earlier days, when cotton was the major reinforcing fibre, adhesion
levels - and performance - were lower: the adhesion obtained between the textile
and the rubber was mechanical, arising from the embedding of the fibre ends into
the rubber matrix. With the advent of the artificial or man-made fibres, these,
being continuous filament yarns, do not shew any significant mechanical adhesion,
so that some form of adhesive treatment is required in order to achieve the
necessary levels of bonding to give the full reinforcing effect.

2.   GENERAL REVIEW OF SYSTEMS

There are four main types of fibre used in the manufacture of the
reinforcing fabrics. These are:-

RAYON: regenerated cellulose
POLYAMIDE: nylons 6.6 and 6,
POLYESTER: poly (ethylene terephthalate),
ARAMID: poly (para-phenylene terephthalamide).

The first of these, rayon, requires a treatment with an adhesive comprising a mixture of a resin with rubber latex. Originally, the resin was casein based, but in 1935, DuPont chemists developed a new system using a resorcinol/formaldehyde resin. This system, a typical formulation of which is given in Table 1, is applied to the fabric, the excess removed, the water dried off and the resultant dip film baked to develop the full adhesive characteristics. A similar resin/latex (RFL) dip system is used for nylon. There are some differences for the optimum results with nylon, but these will be considered in more detail later.

TABLE 1:

BASIC RFL FORMULATION

	PARTS BY WEIGHT	
	DRY	WET
WATER	-	257.8
RESORCINOL	9.4	9.4
FORMALDEHYDE (37% soln.)	5.1	13.8
CAUSTIC SODA (10% soln.)	0.7	7.0
LATEX (40% solids)	84.8	212.0
	100.0	500.0

With polyester, however, these systems do not give adequate adhesion and a pre-treatment is required in order to obtain good adhesion. There are three main types of treatment used with polyester. The first of these is a pre-treatment applied at the spinning stage; various materials have been used but the most important in use at present are epoxy-derived products. An aqueous solution of the epoxy is applied to the yarn just after spinning and a heat treatment given to modify the surface properties of the yarn. After this pre-treatment, a standard RFL is used on the textile construction, applied in much the same manner as for rayon or nylon. The second group of systems use a pre-treatment applied to the textile structure after the assembly process, again followed with an RFL. iso- Cyanate solvent solutions were the first of this type of treatment, but in view of the hazards of such materials are not much used these days; however, a modification of the simple iso-cyanate, the 'blocked' derivative, is used in an aqueous system, again developed by DuPont, known as the 'Shoaf' or D 417 dip. The third range of polyester treatments are modified RFL systems. Basically, the resin component of the dip is modified, to increase the compatibility with the fibre. One of these systems modifies the resin by using a proportion of an aromatic or alicyclic aldehyde, as partial replacement of the formaldehyde, but the most important of these, developed by I.C.I., incorporates a resin containing a proportion of ortho-chlorophenol.

TABLE 2:
RFL MIXING SCHEDULE

PARTS BY WEIGHT

	WET	DRY
WATER	121.8	-
RESORCINOL	9.4	9.4
FORMALDEHYDE (37% soln)	13.8	5.1
CAUSTIC SODA (10% soln)	7.0	0.7
	152.0	15.2
LATEX (40% solids)	212.0	84.8
WATER	136.0	-
RESIN SOLUTION	152.0	15.2
	500.0	100.0

TOTAL SOLIDS CONTENT: 20.0%
RUBBER TO RESIN RATIO: 5.5 :1
FORMALDEHYDE TO RESORCINOL (MOLAR) 2 :1

In a similar way, a pre-treatment is necessary with aramid. Here, a some-what simpler pre-treatment is used, based on an epoxy, followed with an RFL.

It can be seen therefore, that all the textile bonding systems use systems employing an RFL dip, or a modification thereof, so these will be considered in rather more detail.

3. THE R.F.L. SYSTEM

The basic mixing schedule for an R.F.L. dip is shewn in Table 2. This is essentially the same formulation given before, but indicates the order of mixing of the various ingredients. Firstly, the resorcinol is dissolved in water, the formaldehyde added and caustic soda as the catalyst. The resulting resin solution is then added to the diluted latex. This is the basic system; however, there are many aspects of this apparently simple system which have significant effects on the properties of the resultant adhesive.

3.1. THE R.F. RESIN

Firstly, there can be considerable variations in the ratios of the reactants in the resin condensation. The essential chemistry of this condensation is given in Figure 1. The first stage of the reaction is a methylolation of the resorcinol, in the '2' position: this then reacts rapidly with another molecule of resorcinol, giving a dimer. The reaction proceeds in this way, leading to an oligomer generally straight chains of up to about 6 units. After this stage, more complex crosslinking occurs, yielding the final resin product.

The degree of this crosslinking can, of course, be modified by the molar ratios of formaldehyde to resorcinol. Theoretically, a molar ratio of 2:1 should result in a fully crosslinked resin, but generally, in an RFL adhesive, a slightly higher ratio, even perhaps up to 3:1 is used. Surprisingly, however, quite wide variation in this ratio has relatively little effect on the final adhesion obtained with the dip.

Another factor, which does affect the final adhesion is the pH at which the condensation occurs: this is controlled by the addition of alkali (usually sodium hydroxide). This effect can be seen in Fig 2: as the pH, at which the

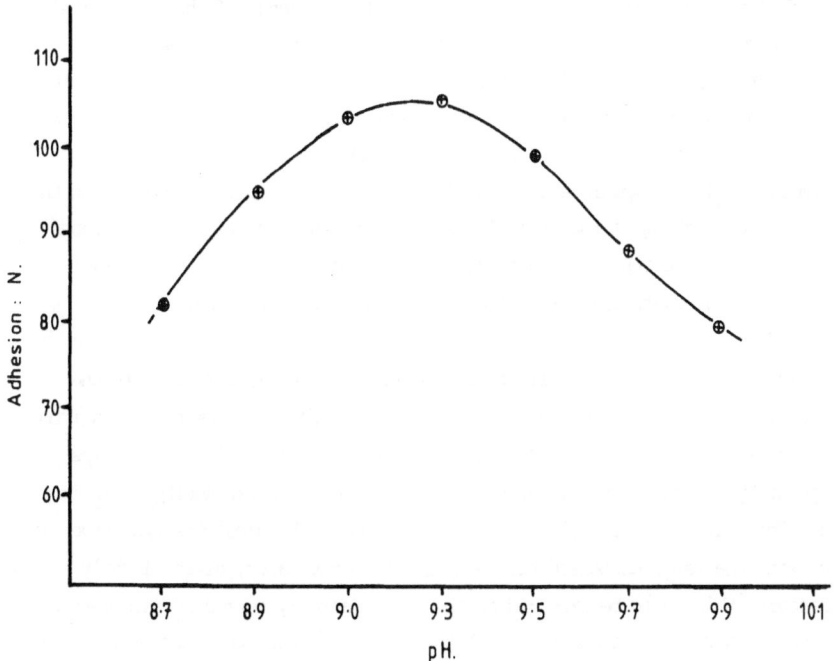

FIGURE 1: CONDENSATION of RESORCINOL with FORMALDEHYDE.

Figure 2: Effect of pH on Adhesion.

dip was prepared, is increased, so the adhesion increases, to an optimum at around pH 9.3, beyond which the adhesion level falls. This optimum pH can vary slightly, depending not only on the actual dip formulation but also on the rubber compound to which the textile is to be bonded, but for most practical applications the exact matching of dip to rubber is of debatable worth, except perhaps where just one fibre type and one compound only are in use.

Earlier, it was stated that the resin solution was added to the latex. Again this simple statement covers various significant effects. In the early days of RFL, when the latex used was ammonia stabilised natural latex, it was essential that the resin be allowed to condense to a reasonably advanced stage before addition to the latex, otherwise the formaldehyde would react preferentially with the ammonia, which both prevents the resin condensation and destabilises the latex. With the advent of the synthetic latices, however, it was found that the resin could be allowed to condense in the presence of the latex. Although the chemical nature of the resultant dips are not significantly different, certain differences in the morphology of the dips and in their behaviour became apparent.

When the resin is allowed to condense in the presence of the latex, the oligomer formed in the earlier stages of the condensation exhibits certain surfactant properties and will displace some of the stabilising soap molecules from around the rubber particles in the latex; this of course leads to a much more intimate dispersion of the resin in the rubber - the two not being thermodynamically compatible and therefore existing in separate phases - as the further crosslinking of the resin leads almost to encapsulation of the rubber particles with the resin, whereas the precondensed resin, already in larger complex molecules, tends to remain as a discrete separate phase.

The main technological differences between the two systems concern the ageing and curing characteristics. In Fig 3 the effects of ageing, timed from the addition of the resin to the latex, on the adhesion, achieved with dips prepared by the two methods, is shewn. Up to about 24 hours ageing (this is the usual "maturation" period allowed in preparation of the dips) the two stage mix, i.e. with the resin condensed before addition to the latex, gives slightly higher adhesion: this is to be expected as the resin, already formed by this method, is the main contributor to adhesion with the textile. After this maturation period, both dips give very similar levels of adhesion, but with the two stage mix, the adhesion tends to fall after about four days, and at least 25% of the optimum adhesion is lost after 14 days, whereas with the single stage mix, there is less than half this loss in adhesion after 42 days storage of the dip.

The other major differences arising from the method of mixing concerns the optimum curing (or baking) conditions for the dip. The cure characteristics of the two systems are shewn in Fig. 4. It can be seen that with the two stage mix, the optimum adhesion is obtained with a baking exposure of around 105 seconds at 150-160°C, whereas the single stage mix gives optimum results with a much shorter exposure at much higher temperatures (around 70 seconds at 205°C)

3.2 THE LATEX COMPONENT

Of course, the resin is only one half of the dip, the latex being the other. Originally, the only latex available was natural, but this was replaced with SBR latex. This worked perfectly satisfactorily with rayon, but still left something to be desired with nylon. This was overcome by the development of a terpolymer latex of styrene, butadiene and 2 - vinyl pyridine. (Generally referred to as VP latex). The replacement of SBR latex with VP latex has a marked effect on the adhesion obtained with nylon, as shewn in Fig. 5. As can be seen, the replacement of around 50% of the SBR with VP gives a very significant increase in adhesion, and further increases are obtained with up to 80% VP. Above this level, there is only a modest increase, but for many applications, it is considered preferable to use 100% VP, to ensure that optimum adhesion is reproducibly obtained.

4. RUBBER COMPOUNDING EFFECTS

This covers the major variables and their effects on the performance of RFL systems: this of course represents only one side of the total bond system, the rubber compound being the other side. There are many more variables in rubber compounding, which can have significant effects on the resultant adhesion.

4.1 POLYMER AND FILLER EFFECTS

The first of these is the polymer. Excluding the speciality elastomers, of the general purpose polymers, SBR generally gives the highest measured adhesion,

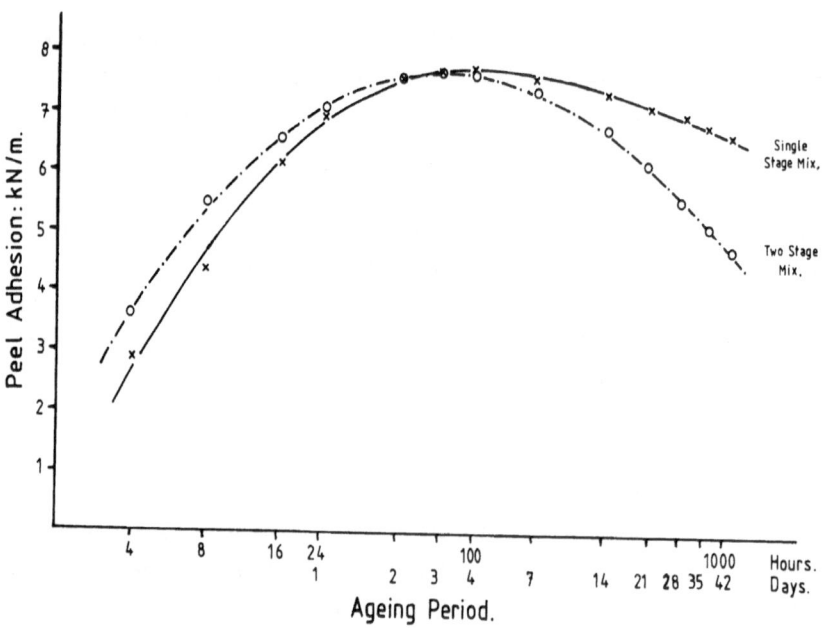

Figure 3: Effects of Ageing on RFL Dips.

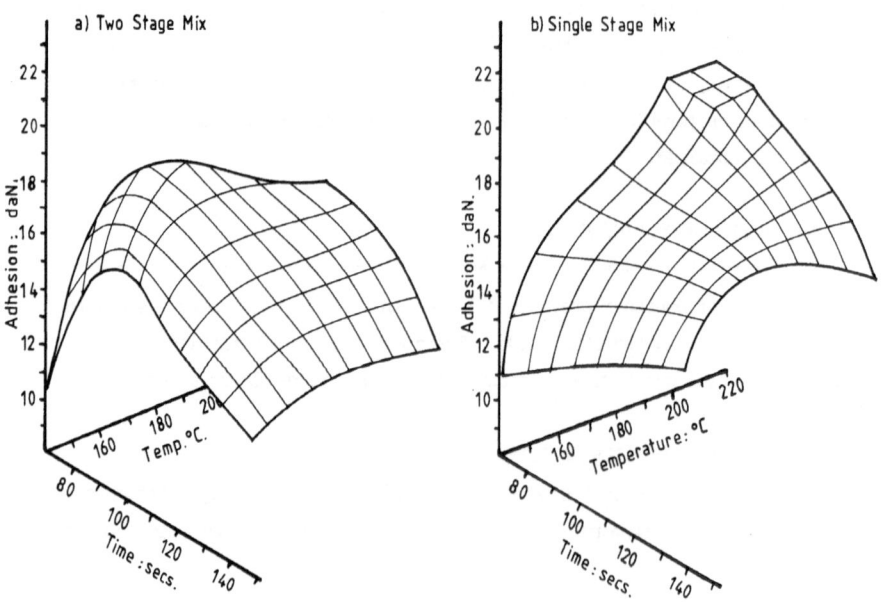

Figure 4: Cure Characteristics of RFL Dips.

with the level dropping as the proportion of natural rubber increases. Secondly, the filler used can have some effects: there are considerable differences in the adhesions obtained between the main classes of fillers, such as the mineral fillers and carbon blacks, but within the different classes, there are not very great differences.

However, for most applications, the polymer/filler system is usually dictated by the properties necessary to meet the service requirements, so that the main variations can only be made in the vulcanizing system.

4.2. THE CHOICE OF CURING SYSTEM

In Table 3 are shewn the adhesion levels obtained with various curing systems. From this it is readily apparent that low sulphur or sulphurless systems do not give good adhesion. In the more conventional systems, those with a slower rate of cure generally give the better adhesions and, where faster cures are required, amine activation of thiazoles (either separately, as with DPG, or "internally" with the sulphenamides) give better adhesions than thiuram or dithiocarbamate activation.

Nevertheless, even though the curing systems chosen gives good adhesion, the actual level of adhesion is still affected by the state of cure achieved in the final composite. Fig. 6 shews these effects. It can be seen that the optimum adhesion is obtained somewhat after the point generally considered to be optimum cure, from the conventional assessment of development of physical properties: however, the adhesion levels tend to drop quite rapidly on continuing cure beyond this optimum level. Comparing this effect of cure time with the development of physical properties, it would appear that the cure for adhesion follows more closely the development of modulus and resilience rather than tear strength, as might have been expected when considering the failure with a peel adhesion test. This suggests, perhaps, that the restriction of deformation at the failure interface, due to the stiffer, higher modulus compound, is more important in achieving high adhesion than are high tear strength or ultimate tensile strength.

TABLE 3

EFFECT OF CURING SYSTEM ON ADHESION

CURING SYSTEM		CURE	PEEL ADHESION *
INGREDIENT	PARTS/100 R.H.C.	MINS/153°C	KN/M
MBTS	0.6		
SULPHUR	2.5	15.0	21.0
CBS	0.5	12.5	18.7
SULPHUR	2.5		
MBTS	0.4		
DPG	0.2	12.5	18.0
SULPHUR	2.5		
NOBS	0.5		
SULPHUR	2.5	15.0	19.1
MBTS	0.4		
TMTD	0.1	10.0	13.7
SULPHUR	2.5		
CBS	4.0		
SULPHUR	0.5	15.0	10.5
CBS	2.0		
SULPHUR	1.0	12.0	12.0
BDTPTS	1.0		
TMTD	3.0	12.0	2.1
CBS	2.0	15.0	1.7
BDTPTS			

MBTS	Di benzthiazyl disulphide
CBS	N -cyclohexyl benzthiazyl sulphenamide
DPG	Di phenyl guanidine
NOBS	N -oxy diethylene benzthiazyl sulphenamide
TMTD	Tetra methyl thiuram disulphide
BDTPTS	Bis (diethyl thio phosphoryl) trisulphide (Vulcadone 3SN: Vulnax Int)
*	2 - ply peel test+ RFL dipped nylon fabric to NR/black compound

11

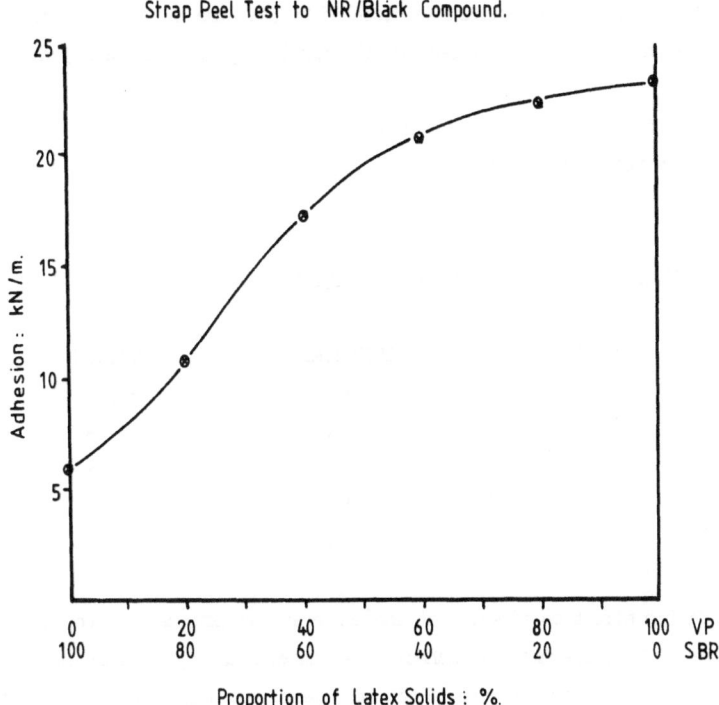

Strap Peel Test to NR/Black Compound.

Figure 5: Effect of Latex Type on Nylon Adhesion.

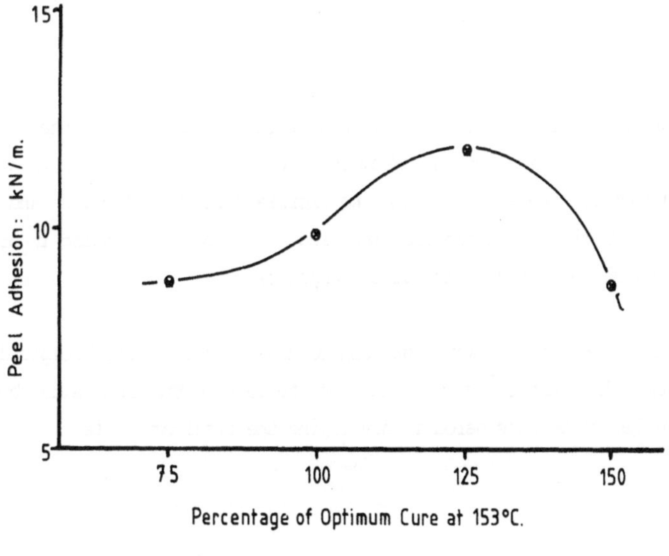

Figure 6: Effect of State of Cure on Adhesion.

Another point to be considered in choosing the cure cycle is the temperature to be used. Often the temperature is dictated by the time necessary to obtain uniform cure throughout the thickness of the final composite, but also, there is an increasing tendency to use higher temperatures, with correspondingly reduced time of cure, to optimize throughput on curing equipment. Table 4 shews the effect of cure temperature (at equivalent state of cure) on adhesion.

TABLE 4

EFFECT OF CURING TEMPERATURE ON ADHESION AT EQUIVALENT CURE.

TEMPERATURE	CURE TIME	ADHESION
°C	Min	kN/m
141	35.0	11.9
153	22.0	11.9
162	11.5	12.1
170	6.0	10.9

There is little effect on adhesion at the temperatures examined; at the highest temperature, 170°C, the measured adhesion is somewhat lower, but this is almost certainly due to the time of cure, 6 minutes, being too short to obtain uniform cure through the test sample, in this case about 8mm thick, rather than to any genuine effect of temperature.

5. CONCLUSION

This briefly covers the main variable factors on both sides of the textile-to-rubber adhesion system. From these various points, it can readily be appreciated that considerable care need be exercised in the selection and finalisation of both the textile treatment and the rubber compounding in order to obtain the optimum adhesion in any given composite.

Generally, there is far more that can be done with the rubber compounding, to "fine-tune" the adhesions system, than on the textile treatment side, but the whole system has to be considered in developing the final composite.

Chapter 2

ASPECTS OF ADHESION PROMOTION WITH FLEXIBLE COATED FABRICS

J.R. HOLKER and A.J.G. SAGAR

Shirley Institute, Didsbury, Manchester M20 8RX

1 INTRODUCTION

Lightweight, flexible coated fabrics are widely used in civil and military outdoor apparel, protective covers and workwear, hand-luggage, upholstery, and tentage. The base fabrics are woven or knitted from staple-spun yarns, or from multifilament synthetic yarns, and their surfaces may be made more receptive by raising or by texturing the filament yarns; some nonwoven fabrics are also used.

PVC still accounts for the largest share of the market for coating compounds, despite the decline in its use for automotive upholstery, at one time a major outlet. Polyurethanes (PUs) are more expensive than PVC, and conditions of application are probably more critical, but their excellent mechanical properties, durability at considerably lower coating weights, and aesthetic appeal make them the preferred coatings for lightweight rainwear, military combat uniforms, small tents, and some soft hand-luggage. The aesthetic advantages of PUs have also forced natural and synthetic rubber coatings into uses where handle and appearance are of lesser importance, e.g. protective clothing of various kinds.

Heavier coatings of PVC plastisols and rubber doughs may be applied by calendering, and special hot-melt processes are also available for thermoplastic PUs. Lightweight coatings of PVC and of solvent-based PUs and rubbers, however, are most simply and economically applied by the direct knife-on-roll or knife-on-air method; the desired weight of coating can be built up, if necessary, over several passes. For practical reasons, this method is restricted to closely-woven or closely-knitted, smooth, dimensionally stable fabric substrates. Unstable, open and rough-surfaced fabrics must be coated indirectly, either by transfer-coating from a release paper or by film lamination with a separate adhesive.

Wake[1] has discussed the factors implicated in achieving good adhesion between polymer coatings and a textile substrate and in maintaining the bond under wet conditions. Referring specifically to the high-speed application of lightweight coatings, he stresses that some increase in specific adhesion, coupled with the use of polymer solutions of sufficiently low viscosity to ensure adequate penetration of the fabric, are essential to the formation of strong, durable bonds. For some products, however, notably apparel, the latter requirement conflicts with the need to restrict penetration in order to retain optimal softness, handle, and tear strength. Transfer coating of open or rough-surfaced fabrics, of course, affords ample scope for creating a good mechanical bond with minimal penetration, but in the direct coating of tightly woven fabrics the coating must be forced into the yarn interstices and around the surface fibres to ensure adequate adhesion; in the latter case, therefore, a balance has to be struck between adhesion and penetration into (and consequent stiffening of) the fabric.

Experience indicates that there is scope for improvement in this balance of aesthetic and physical properties against adhesion of lightweight coatings, and also that delamination in service as the result of repeated flexing, particularly in cold conditions, is still a matter for concern. Any practicable methods that lead to an increase in specific adhesion should, therefore, be of considerable value to the coating industry. Treatments to improve adhesion to fabrics for heavy-duty uses, such as beltings and hose, and to tyre-cord are routinely used in industry, and have been thoroughly reviewed by Wootton[2], but generally speaking they are not suitable for lightweight materials. Some attempts to effect improvements, so far directed almost exclusively to the PU coating of lightweight nylon fabric are now recorded; they may, of course, also have relevance to heavier fabrics.

2 MODIFICATION OF NYLON 6.6 FABRICS

A prerequisite of the search for a treatment that could be applied to the surface of polyamide fibres, in order to introduce functionality that might promote better adhesion of coating polymers, was that the treatment should be readily applicable on conventional textile processing equipment and not involve economically or environmentally unacceptable reagents. Furthermore, any process implicating the fixation of substantial amounts of resin to fibre surfaces should be avoided, because of the dangers of inter-fibre bonding, with its consequent stiffening effect and likely adverse influence on tear strength.

One known treatment that had previously been exploited at the Shirley Institute[3,4] and elsewhere[5], as a means of imparting better soil-release characteristics to nylon, is the fixation of polymeric carboxylic acids at elevated temperature. The same end can be achieved, albeit more expensively, by the graft polymer-ization of acrylic acid and related monomers, initiated chemically[6] or by high energy radiation[7]; indeed, the latter technique has been reported to improve the adhesion between nylon and a nitrile rubber[8].

The reaction of poly(acrylic acid) (PAA) with nylon was first studied by Nuessle and Kine[9], who reported the strong specific adhesion between these materials, and by Nuessle and Crawford[10], who exploited it to deliberately stiffen fabrics. The unwanted fixation of the acid through overzealous drying has frequently caused problems for finishers of woven nylon fabrics, for which it is a preferred warp-sizing agent. The mechanism of fixation of PAA to nylon has never been conclusively demonstrated; acylation of amine end-groups, transamidation, and imide formation have all been suggested[11]. Certainly, it does not seem to be ascribable simply to intermolecular anhydride formation, since there is only a small loss in carboxyl content on curing, the finish is resistant to extraction with hot alkali, and it is not durable on other fibres (acrylic, polyester, cellulose acetate) Furthermore, at the lowest temperatures needed to effect the reaction, dehydration of PAA is very slow, being incomplete after many hours at $170^{\circ}C$[11]. We have observed no significant weight loss below $200^{\circ}C$ by thermogravimetry, whilst extraction with a solvent for nylon, e.g. 90% formic acid, left a thin envelope of insoluble material.

3 POLYURETHANE COATING OF PAA-MODIFIED NYLON

3.1 Effect of PAA Concentration

As soil-release agents, several of the polymeric acids were still effective after at least 20 laundering cycles at $60^{\circ}C$, when applied from 1% aqueous solutions (actual fixation was much lower). Concentrations of this order were considered too high for the present purpose for two reasons:

1. Resistance to frequently repeated laundering is not necessary
2. Because PAA itself yields rather stiff, brittle films, a heavy coating could be a potential source of cohesive failure.

A first step in the investigation, therefore, was to define the levels of application of PAA. A scoured, dyed, and heat-set filament-woven nylon (ca. 120 $g.m^{-2}$), of a type widely used in both civilian and military garments, was selected as a suitable base fabric. This fabric was impregnated with solutions of PAA of molecular weight ca. 230000 (0.025 to 0.2% w/v) containing a non-ionic detergent, Synperonic NX (0.05% w/v) to ensure even wetting-out. Fixation was achieved by drying and curing the fabrics in a single pass through a forced-draught oven at $190^{\circ}C$ for 2.5 min; lengths of unmodified fabric and of fabric treated with surfactant only were given the same thermal treatment. The fabrics were then coated with three different two-component polyurethanes, two of commercial provenance, by the direct knife-on-roll method, typically with a knife-gap of 150 microns, on a pilot-scale coating machine.

The coating polymers were:

1. Impranil C + Imprafix TH (13% w/w solution on Impranil C solids) as cross-linking agent in methyl ethyl ketone (MEK).

2. Witcoflex 635 + Witcobond BY98 (12% w/w solution on Witcoflex 635 solution) in ethyl acetate.

   Witcobond BY98 is said to give a softer and drier coating with greater resistance to hydrolytic ageing than Imprafix TH or its equivalent.

3. A polymer prepared from 4,4'-diisocyanato-dicyclohexylmethane (HMDI; Desmodur W), butane - 1,4-diol, and polycaprolactone diol (MW 1250) in the molar ratio 2:1:1, by dropwise addition of the isocyanate to the mixed diols. Imprafix TH (13%) was again the cross-linking agent and MEK the solvent.

Table 1:

Effect of PAA concentration

PAA(%)	Coating Wt. (g/m^2)	Peel Strength (N/5cm)	Tear Strength (N)		H.S. Head (cm) Flexed	
			W	F	W	F
IMPRANIL C						
-	30	60	67	58	36	23
surfactant only	34	74	66	59	51	62
0.025	38	97	43	46	36	30
0.05	39	100+	38	39	28	21
0.10	33	98	56	49	10	15
WITCOFLEX 635						
-	50	34	52	60	150+	150+
surfactant only	47	33	60	70	"	"
0.025	47	39	59	67	"	"
0.05	46	62	48	55	"	"
0.10	50	63	38	40	"	"
POLYMER A						
-	28	65	63	53	20	24
0.10	21	100+	42	37	6	4

Note: Fabrics were tested in accordance with B.S. 3424: Methods of Test for Coated Fabrics; W = warp direction, F = fill (weft) direction.

Unfortunately, the scale of the experiments did not allow sufficient time for coating weights to be adjusted accurately; the usual value for a fabric of this type would be ca. 30 g m^{-2}.

The sharp improvements in peel strength of all three coatings at quite low levels of modification with PAA are clearly seen in Table 1. For comparison, a typical commercially coated fabric at 33 gm^{-2} coating weight gave a value of 56 N/5cm. The civilian and military specifications for this type of fabric (BS 3546, Part I, 1981, and Ministry of Defence SC 3418G) demand a minimum of 35 N/5cm. It is of passing interest that some improvement in adhesion of the Impranil C coating occurred when the fabric was merely treated with Synperonic NX; the ability of a nylon 6.6 fabric to retain a non-ionic surfactant (Triton X-100) applied at high temperature (100-175°C) is reflected in a greatly increased wettability and modified dyeability. No such improvement was observed with the Witcoflex 635 coating. Peel strengths were generally lower for Witcoflex 635 than for Impranil C, so direct comparison of the two systems is not possible; however, the Witcoflex 635 coating appeared to require a higher level of fabric activation.

Improvements in adhesion of flexible coatings can always be achieved by forcing the coating compound deeper into the fabric, but this method has severe limitations, since it leads to a marked deterioration in fabric handle (stiffness) and tear strength; it also requires a heavier loading of polymer if crowns of the yarns are not to be starved of coating. The sensitivity of tear strength to coating weight is seen in a comparison of control and PAA treated samples coated with two weights of Impranil C + Imprafix TH (Table 2).

Table 2:
Effect of coating weight on tear strength

Fabric	Coating Wt. (g/m^2)	Tear Strength (N)	
		Warp	Weft
Control	30	67	58
	41	44	51
PAA-treated (0.1% w/v)	33	56	49
	42	31	34

Available data are, as yet, insufficient to establish any clear quantitative relationship between the level of PAA treatment and tear strength of the coated fabric, but the trend was clearly downwards for all three polymers. Electron micrographs show that the depth of penetration of coatings was no greater on the PAA-treated fabric than on the controls; the lowering of tear strength must, therefore, be related to the firmer anchoring of the fibres and its consequent effect on stress dissipation.

For practical purposes, it was concluded that no further benefit was to be gained in respect of peel adhesion from treatment of nylon with PAA at concentrations above 0.1%. Even at this level, improved adhesion was accompanied by some loss in tear strength. For the fabric in question, the specifications call for a tear strength minimum of 90N in both warp and weft directions, but in our experience commercially coated fabrics frequently fail to meet this requirement. One of the most severe tests for polyurethane coated fabrics for outerwear is their ability to retain a high resistance to penetration by water (BS 3424, Method 29C) after repeated rapid flexing (200,000 cycles, Schildknecht); flexing may be preceded by accelerated humidity ageing to probe any susceptibility to hydrolytic degradation in use or storage, a factor of importance in tropical climates. Catastrophic failure of the coating manifests itself in localised delamination and cracking along the flexing lines. Good adhesion obviously contributes to minimising the problem, but the elastic properties of the coating also play an important, perhaps dominant, role. Under rapid cyclic loading, as in the Schildknecht test, a stiff film with inadequate elastic recovery will suffer progressive hysteresis losses , culminating in rupture.

The values of hydrostatic head resistance after flexing (no ageing) given in Table 1 demonstrate, in the case of the Impranil C coating, that improved adhesion led to no improvement; the pass value, incidentally, is 70cm. They also underline the difference between Impranil C and the softer Witcoflex 635; the latter coating retained its high resistance to water penetration, irrespective of the level of PAA treatment, whereas the Impranil C coatings all responded poorly. A commercially coated fabric also

behaved indifferently, with a hydrostatic head of only 45cm after
flexing. Failure of the experimental coating (Polymer A) was not
unexpected; tensile behaviour of the film indicated that elastic
recovery would be poor.

One consequence of the foregoing observations is that the
better adhesion to PAA-treated nylon fabric might best be
exploited by keeping penetration of the coating into the fabric
as low as possible, consistent with maintaining peel strength
at an adequate level, rather than aiming simply to achieve maximal
adhesion. In this way, better retention of tear strength might
result.

Superior adhesion of the PU coatings to PAA-modified
nylon could be explained by the formation of covalent bonds
between the polymer and carboxyl groups on the fibre surfaces,
through mutual reaction with the cross-linking agent. However,
a one-component fully extended PU coating (Witcoflex 303), where
no cross-linking agent is involved, was also found to exhibit
better adhesion to the modified fabric. Peel strength rose
from a mere 13N/5cm for the untreated fabric to 25N/5cm, with
only a modest fall in tear strength, at a coating weight of
ca. 30 gm^{-2}.

3.2   Other Polymeric Acids

When it had been amply and repeatedly demonstrated that
pretreatment with PAA resulted in significantly improved adhesion
to PU coatings, the study was extended to other polymeric acids,
viz polymethacrylic acid, polyitaconic acid, and the alternating
copolymers of maleic acid with ethylene and with methyl vinyl
ether. They were applied under the same conditions, but none
of them proved to have any beneficial effect on adhesion.
Treatment of nylon with the involatile pyromellitic acid
(benzene -1,2,4,5-tetracarboxylic acid) proved equally ineffective,
although alkalimetry showed that some reaction with the fibre
had occurred.

3.3 Wet Adhesion

For many purposes it is essential that fabric coatings retain good adhesion in the wet condition; indeed, both BS 3546 Pt.I and SC 3418G demand the same minimum value (35N/5cm) for both dry and wet peel strength. Some loss of peel strength is usually experienced on wetting, particularly if the fibre or coating is sensitive to swelling in water, although this loss may be wholly or substantially recovered when the fabric is dried. SC 3418G, in fact, also lays down the same minimum peel strength after a standard laundering. Losses in adhesion may also occur, often more seriously, in dry-cleaning solvents as a result of swelling of the coating; as protection against inadequate solvent resistance, both the civilian and military specifications again insist on a minimum peel strength of 35N/5cm after dry-cleaning (not whilst wet with solvent).

In the preferred method of test for wet peel adhesion, the samples are not immersed in water, but in 2% sodium oleate solution,for 30 min at ca.20°C. This method has been adopted in order to obviate drying-out of the samples before testing is complete.

Adhesion of some commercial coating polymers proved to be more sensitive to wetting for PAA-treated nylon than for untreated fabric, but even in the wet state the treated fabric remained superior (Table 3).

The drop in peel strength was greatest with the Impranil C coating, which had responded the most favourably to PAA treatment in respect of dry adhesion, and least with the Witcoflex 635. That this effect is largely temporary is shown by the recovery of adhesion to PAA-treated fabric of both Impranil C and CHW coatings after a wetting cycle.

3.4 Effect of PAA Treatment on Stiffness and Liveliness

Attempts to monitor the effect of changes in adhesion, as the result of fabric pretreatment, on stiffness and handle (liveliness) were made, using the Shirley Bending Hysteresis Tester, illustrated diagrammatically in Fig.1.

Table 3:

Wet peel strength of PU coatings

PAA-treated	Coating Wt. (g/m^2)	Adhesion (N/5cm)			
		Dry	Oleate	Water	Redried
IMPRANIL C					
No	32	39	34	40	39
Yes	31	100	59	-	91
IMPRANIL CHW					
No	36	36	27	-	6
Yes	43	68	48	49	73
WITCOFLEX 635					
No	36	47	40	-	-
Yes	45	57	48	-	-
IMPRANIL CLS-02					
No	35	NOT MEASURABLE			
Yes	37				
COMMERCIAL FABRIC					
No	33	56	47	51	50

In this instrument a small fabric specimen is taken through a bending cycle; that is, it is bent slowly first in one direction and then in the other. The variation of the bending couple (resistance to bending) with the curvature of the specimen can be recorded throughout the cycle.

The instrument is attached to an Instron Tensile Tester. The specimen AB is held in a rotating grip C, which is driven from the Instron crosshead by means of a steel tape and a pulley; the other end is held in a grip which forms part of a very light arm. The lower end of the arm is a free fit between two pins attached to a force transducer, whose output is proportional to the bending couple of the specimen and can be fed to the Instron pen-recorder. By taking the specimen through a bending cycle

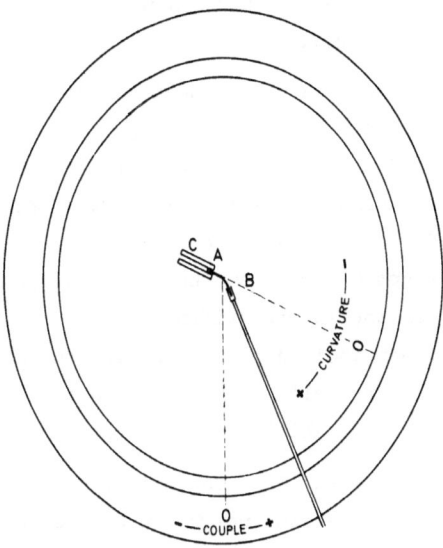

Fig.1   Principle of bending-hysteresis tester

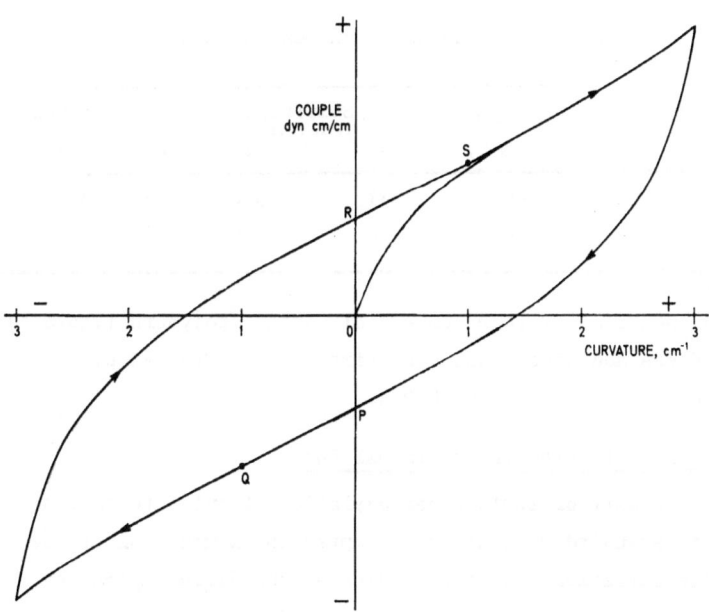

Fig.2   Typical bending-hysteresis curve

(usually $\pm$ $3cm^{-1}$) a bending-hysteresis loop is plotted by the pen recorder as shown in Fig.2.

Two important quantities may be derived from the bending hysteresis loop. The first, which is the slope of straight portions of the curve QR and RS, is the elastic flexural rigidity, Go; this represents the purely elastic component of fabric stiffness. The second quantity is the coercive couple, Co, which is given by the intercept OP or OR on the couple axis, and represents that part of the fabric stiffness that arises from frictional forces between individual fibres. The sum Go + Co represents the stiffness as felt by the hand and the ratio Go/Co is related to liveliness, or ability to recover from gentle bending.

The method demonstrates, e.g,,the differences between Witcoflex 635 coatings at the same weight, when cross-linked with two different agents, Imprafix TH and Witcobond BY98, the latter compound yielding the softer and more lively product (Table 4).

Table 4:

Effect of cross-linking agent on bending hysteresis

Cross-linker	Coating Wt. (g/m^2)	Stiffness (dyne.cm)		Liveliness	
		W	F	W	F
Imprafix TH	45	341	350	3.5	3.4
Witcobond BY98	46	240	196	5.0	4.0

In the trial of four commercial coating polymers (Table 5) both stiffness and liveliness were virtually unaffected by pretreatment of the fabric with PAA.

3.5   Bonding of Nylon to Functional PUs

The ability of surface carboxylation of nylon to promote adhesion to standard PU coatings prompted an examination of the alternative situation, ie. the bonding of unmodified nylon to polymers carrying functional groups. Accordingly, a group of polymers was synthesised, based on a control identical with

Table 5:

Effect of PAA pretreatment on tear strength and bending
hysteresis

PAA-treated	Coating Wt. (g/m^2)	Tear Strength(N) W	F	Stiffness (dyne.cm)	Liveliness
IMPRANIL C					
No	32	73	80	356	3.8
Yes	31	58	62	420	4.6
IMPRANIL CHW					
No	36	89	98	270	3.1
Yes	43	78	79	242	3.4
WITCOFLEX 635					
No	36	85	96	368	3.1
Yes	45	72	82	345	3.5
IMPRANIL CLS-02					
No	35	94	103	303	3.8
Yes	37	123	120	246	3.9
COMMERCIAL FABRIC					
No	33	69	72	235	5.2

polymer A (Table 1), i.e. HMDI:butane -1,4 - diol (TMG):
polycaprolactone diol (MW 1250) in the molar ratio 2:1:1.
In the variants, 0.25mol of the TMG was replaced by bishydroxy-
methylpropionic acid $(HOCH_2)_2CMeCOOH$, N-methyldiethanolamine (NMDEA)
or the diethanolamine derived from a mixture of primary amines
obtained by reductive amination of the fatty acids of coconut
oil (Ethomeen C12). The polymers gave strong unsupported films
(Table 6), which in the case of the Ethomeen-based variant is
somewhat surprising; a high level of substitution of a component
carrying a bulky side-chain was expected to interfere
with intermolecular hydrogen bonding. When coated onto the usual
nylon fabric, the functionally modified polymers all had better
dry adhesion than the control polymer. Carboxylated PVC plastisols

have been reported to have better adhesion to fabric than the
homopolymer[13].

Table 6:

Film properties and adhesion to nylon of functional PUs

POLYMER	Limiting Visc.No. $(dl.g^{-1})$	Film Strength (MNm^{-2})	Elong. (%)	Coating Wt. (g/m^2)	Peel Strength (N/5cm)
CONTROL	0.44	36.7	390	28	40
+COOH	0.30	33.4	450	29	57
+NMDEA	0.34	41.9	480	35	66
+ETHOMEEN C12	0.38	39.5	550	36	63

3.6   Coating of PAA-nylon with Basic PUs

A further series of experiments was then designed to test
whether or not simultaneous modification of the nylon with PAA
and of the coating polymer with a basic comonomer (N-methyldiethanolamine, NMDEA) would lead to a synergistic effect on
adhesion, through formation of salt-like cross-links between
the adherends. Five polymers were prepared by slow addition of
isophorone diisocyanate (2 mol) to a mixture of butane-1,4-diol
(1-x mol, where x = 0, 0.25, 0.5, 0.75 and 1.0). N-methyldiethanolamine (x mol), and polycaprolactone diol (MW2000)(1 mol)
in boiling methyl ethyl ketone. They were then coated onto
untreated and PAA-treated nylon fabric in the usual way, again
with Imprafix TH (13% w/w solution/solids) as the cross-linking
agent. Dry peel strengths (Table 7) were in every case higher
on the treated nylon, but contrary to expectation no additional
benefit accrued from insertion of the basic comonomers into the
PU structures; even at an NMDEA content of 1.0 mol, dry adhesion
was lower than with the unmodified polymer. One feature of the
behaviour of the PAA-nylon + basic copolymer system that requires
closer examination is that wet adhesion, whether in water alone
or in 2% sodium oleate solution, was significantly higher than
dry adhesion for the polymer containing 0.25 mol NMDEA, but not

Table 7:
Adhesion of basic PUs to PAA-treated nylon

NMDEA (mol.%)	Coating Wt$_2$ (g/m^2)	Adhesion (N/5cm)			
		Dry	Oleate	Water	Redried
		Untreated nylon			
0	26	21 (38)	20	27	(50)
25	32	26	29	28	(41)
50	29	26	27	22	35
75	22	26	28	34	39
100	31	25	23	22	38
		PAA-treated nylon			
0	27	53 (69)	61	60	(66)
25	29	32	50	49	58
50	31	36	38	33	43
75	28	37	37	31	43
100	33	45	35	36	88

Note: Values in brackets are for samples tested several weeks
       after preparation.

for the other members of the series. Moreover, when all the
water-soaked samples were re-dried, their dry peel adhesion was
found to be greater than the initial values - by almost 100%
where NMDEA had completely replaced the butane-1,4-diol. It is
tempting to invoke increased molecular mobility at the interface,
resulting in the creation of new salt cross-links in an aqueous
environment. Films of the NMDEA-modified polymers, incidentally,
adhered very strongly to glass, even when wet; they could only
be removed by soaking in dilute acid, thus implying some polar
interaction between the adherends.

The composition of these polymers had little or no effect
on tear strengths, but again values were generally lower on PAA-
treated fabric than on untreated fabric (Table 8). There was also
no obvious pattern of change in either stiffness or liveliness
with composition or with treated/untreated fabric.

Table 8:

Physical properties of basic PU-coated nylons

NMDEA (mol.%)	Coating Wt. (g/m^2)	Tear Strength(N) W	F	Stiffness (dyne.cm)	Liveliness
		Untreated nylon			
0	26	75	76	466	4.1
25	32	76	99	552	4.3
50	29	79	87	534	4.3
75	22	86	91	557	4.6
100	31	81	92	410	3.6
		PAA-treated nylon			
0	27	70	80	564	4.1
25	29	71	83	528	3.9
50	31	71	83	575	4.5
75	28	78	89	617	4.4
100	33	75	87	390	3.5

3.7 Commercial Coating of PAA-treated Nylon

The results obtained in the laboratory, demonstrating repeatedly the improvements in adhesion of PU coatings to PAA-modified nylon, were considered sufficiently encouraging to warrant a short trial under commercial coating conditions. The fabric, a dyed, heat-set filament woven nylon (50m, 1.5m width) similar to that used in the laboratory work, was treated with 0.075% PAA solution in the usual way. It was then commission-coated with a commercial polymer; a similar length of untreated fabric was also coated as a control. From the test results (Table 9), this trial followed the expected pattern, although coating weights were rather too low. Peel strengths were all significantly better on the PAA-treated fabric (dry, wet, redried, and dry-cleaned). Both fabrics failed the flexing test, perhaps as a consequence of the low coating weight; the untreated fabric withstood abrasion and dry-cleaning better, in respect of retained hydrostatic head resistance. Tear strength loss was slightly greater on the treated fabric, stiffness was about the same, and the treated fabric was rather more lively.

Table 9:

Commercial coating of PAA-treated nylon.

	Untreated nylon	PAA-treated nylon
Coating (g/m^2)	20 (33)	20 (29)
Peel strength (N/5cm)		
Dry	27 (40)	53 (89)
Wet (Na oleate)	15	37
Dry-cleaned	21	48
Washed	30	52
Hydrostatic head (cm)		
As received	150+	108
Flexed	46(W), 26(F)	38
Abraded	73	54
Dry-cleaned	150+	94
Tear strength (N)		
Warp	63 (61)	53 (54)
Weft	76 (71)	63 (64)
Stiffness (dyne.cm)	440 (330)	454(W),854(F) (335/575)
Liveliness	3.2 (3.2)	4.1 (3.6)

Note: Values in brackets were obtained on laboratory-coated fabric (Impranil C)

4   OTHER POLYMER-NYLON COMBINATIONS

4.1   PVC Coating of PAA-treated Nylon

The first results from a cursory examination indicate that some improvement in peel adhesion can also be achieved in this system (Table 10). Fabric treated with PAA (0.1% w/v) in the usual way was direct-coated (knife-on-roll) with two PVC plastisols, one of which was pigmented and of distinctly higher viscosity than the clear sample. In each case, dry peel adhesion increased by some 30N/5cm as a consequence of the PAA treatment, and there was no significant alteration in tear strength.

Table 10:

PVC coating of PAA-treated nylon

Coating	Pigmented PVC		Unpigmented PVC	
PAA treatment	No	Yes	No	Yes
Coating wt. (g/m^2)	180	173	134	130
Dry peel strength (N/5cm)	22	52	15	45
Tear strength (N)				
Warp	89	82	97	106
Weft	102	95	118	117

4.2 Coating of Nylon Pretreated with a PU Dispersion

A nylon fabric of the type used throughout this work was impregnated with a very dilute PU dispersion (Witcobond W234, 1g/1) and dried at 90°C for 4-5 min. to coagulate the polymer and thus impart a thin, water-resistant coating to the fibres. When coated with the same clear PVC plastisol as in Section 4.1, the pretreated fabric again showed a distinct improvement in dry adhesion (Table 11), with no adverse effect on tear strength.

In the one experiment conducted to date, the same treatment was also very effective in promoting the adhesion of an Impranil C + Imprafix TH coating. A peel strength comparable with the best observed for PAA-treated nylon was achieved, although there was a noticeable sacrifice in tear strength (Table 11). Witcobond W234 forms rather hard films, a fact which might account in part for this loss in tear strength and also for the subjectively observed stiffening; a softer polymer will probably give better results.

Table 11 :

Coating of PU-treated nylon

Coating	Unpigmented PVC		Impranil C	
PU treated	No	Yes	No	Yes
Coating wt. (g/m^2)	134	147	32	39
Peel strength $(N/5cm)$				
Dry	15	50	39	109
Wet, Na oleate	-	-	34	85
Tear strength (N)				
Warp	97	95	73	46
Weft	118	123	80	55

ACKNOWLEDGEMENTS

The work reported here forms part of a continuing programme of research into improvements in coated fabrics. We wish to thank the Department of Trade and Industry for supporting the studies through its Textile and Other Manufactures Requirements Board, and Mr. J. Holmes of Gordon and Fairclough Ltd. for arranging the commercial coating trial.

REFERENCES

1. Wake, W.C 'The adhesion of polymers to fibrous masses', J. Coated Fabrics, vol.3, 84-95, Oct. 1973.

2. Wootton, D.B. 'The present position of tyre cord adhesives', Developments in Adhesives - 1, Applied Science Publishers,1977.

3. Holker, J.R. 'Polyacrylic acid - a soil-release finish for nylon', Textile Institute and Industry, vol.8, No.11, 305-307, Nov. 1970.

4. Holker, J.R. 'Polymeric acids and cellulose acetate as soil-release finishes', Shirley Institute Bulletin, vol.46, 124-131, Oct. 1973.

5. Moyse, J.A. 'Finishing processes for synthetic and blended fibre textiles to confer soil release and related effects', Textilveredlung, vol.5, No.5, 377-385, 1970.

6. Haworth, S and Holker, J.R. 'Cerium initiated polymerization of some vinyl compounds in polyamide fibres', J. Soc. Dyers Colourists, vol. 82, No.7, 257-264, 1966.

7. Magat, E.E, Miller, J.K, Tanner, D, and Zimmerman, J. 'Grafts of nylon and unsaturated acids', J. Polymer Sci., Part C, no.4, 615-629, 1963.

8. Ryabchikova, G.G, Kabanov, V. Ya, Khaimina, A.A and Baranets, Yu.N. 'Modification of polamide fibre by radiation/chemical grafting to increase its adhesion to rubber', Kauch.i Rezina, No. 1, 37-39, 1975.

9. Nuessle, A.C and Kine, B.B. 'Acrylic resins in textile processing', Ind. Eng. Chem., vol. 45, 1287-1293, 1953.

10. Nuessle, A.C and Crawford, R.F 'The reaction of polyacrylic acid with nylon', Text Res.J, vol.23, No.7, 462-468, July 1953.

11. Eisenberg, A, Yokoyama, T. and Sambalido, E. 'Dehydration kinetics and glass transition of poly(acrylic acid)', J. Polymer Sci., Part A-1, vol.7, 1717 - 1728, 1969.

12. Needles, H.L, Wen-chu Lu, Alger, K and Varma, D.S. 'Effect of nonionic surfactant and heat on selected properties of nylon 6.6', J. Appl. Polymer Sci., vol.25, 1745 - 1753, 1980.

13. Russell, J.R and Ward, D.W. 'Reactive functionality in dispersion resins - adhesion of carboxyl-modified plastisols to fabrics', Symposium on Coated Fabrics Technology, 30-39, 1973. American Association of Textile Chemists and Colorists.

Chapter 3

THE ADHESION OF SOME ETHYLENE-VINYL ACETATE COPOLYMERS APPLIED AS HOT MELT COATINGS TO METALS.

T A HATZINIKOLAOU and D E PACKHAM

School of Materials Science, University of Bath

1. INTRODUCTION

Hot melt adhesive systems have advantages in many bonding and coating applications. They are free of solvents which can cause fire or environmental damage, they may be stored for long periods, and the bond forms rapidly on the application of heat. Ethylene-vinylacetate (EVA) copolymers are an important component of many hot melt systems, contributing significantly to their adhesion[1].

The practical adhesion of a polymer to a substrate is the result of a complex interaction of many factors. The physical and chemical state of the interface and interactions which occur there have an influence, as does the mechanical response to stress of the polymer itself. Thus in hot melt systems based on polyethylene the possibility of surface oxidation, the surface topography of the substrate and the yield and ultimate properties of the polymer can all conspire to determine the measured adhesion[2]. Good practical adhesion is obtained where considerable dissipation of energy occurs within the polymer when the coating is peeled. The mechanism by which this dissipation occurs appears to be different depending on whether the substrate surface is conventionally prepared, giving a fairly smooth surface, or prepared to give a microfibrous topography[3].

Vinyl acetate may be copolymerised with ethylene by a radical mechanism to provide an interesting series of copolymers. As the vinyl acetate content increases the crystallinity of the polymer decreases, and the polarity increases. Thus the mechanical properties change progressively, for example yield strength and modulus fall, but impact energy rises, as vinyl acetate content increases[4].

This paper reports results from a current project in which the adhesion of EVAs applied to metals as hot melt coatings is being investigated. Four copolymers have been selected with approximately constant molecular weight, but varying vinyl acetate content. Conventionally prepared copper and steel substrates are used together with copper oxidised in an alkaline chlorite solution to give a microfibrous copper (II) oxide surface[3]. The magnitude of adhesion measured is discussed in terms of the possible contribution of bulk and surface energy loss processes.

2. EXPERIMENTAL METHODS

2.1 Materials

Two metals were used, steel and copper. The steel was general purpose mild steel sheet 1.26 mm thick (BS1149) supplied by Woodberry Chillcott, Bristol and the copper was a deoxidised sheet (BS1172) 1.2 mm thick supplied by H Righton, Bristol. Both metals were cut to panels 10 cm by 15 cm. The EVA copolymers used were supplied in powder form, three by ICI, referred to as A, B and C and the fourth (polymer "D") by du Pont. Some of their properties were measured and the results (Table 1) are very similar to those quoted by the manufacturers.

Table 1

The ethylene vinyl acetate copolymers used.

Designation in this paper	D	C	B	A
Manufacturer's code	Elvax 750	554–080	540–041	2805–042
Melt flow index (g/10 min)	6.2	3.1	7.2	7.2
Percent vinyl acetate by wt	9.8	12.0	17.5	27.5
Percent crystallinity	47.0	46.0	38.2	22.0

2.2 Substrate pretreatment

2.2.1 Etched steel The steel panels were degreased in boiling trichloroethane for 20 minutes and then etched in 5 M hydrochloric acid for 30 seconds, followed by rinsing with water and acetone.

2.2.2 Chemically polished copper The copper panels were immersed in dilute hydrochloric acid and rinsed in distilled water and then acetone. They were next degreased in boiling trichloroethane for 20 minutes and immersed for 10 minutes in a polishing solution consisting of:

> 60 ml orthophosphoric acid (SG 1.75)
>
> 10 ml nitric acid (SG 1.42)
>
> 30 ml acetic anhydride. and
>
> 8 ml distilled water.

Finally. they were rinsed with water and acetone.

2.2.3 Chlorite formed films on copper Chemically polished samples prepared as in 2.2.2 were immersed for 20 minutes at $90^{o}C$ in a solution containing:

> 30 g/l sodium chlorite
>
> 100 g/l tri-sodium orthophosphate and.
>
> 50 g/l sodium hydroxide.

2.3 Preparation of coatings

The metal panels were preheated for 10 minutes at $200^{o}C$ in an oven under vacuum of about 150 torr. On removal from the oven. polymer powder was sprinkled on the metals. the excess powder was removed. and the samples were replaced in the oven for a 20 minute coating period at $200^{o}C$. For the first six minutes the coating time were again under vacuum to avoid the formation of air bubbles at the interface. and then the atmospheric pressure was restored.

After coating the specimens were removed from the oven and were left to cool on the bench.

Some coatings were prepared with an essentially inextensible. non-dissipative fabric backing (loomstate cotton duck to BS 4F 55). The procedure was the same as that described above, except that the coating period was 17 minutes at the end of which the backing was placed on top of the polymer, and the sample placed between the platterns of a press at $200^{\circ}C$. It was pressed for three minutes using spacers to obtained the required thickness.

2.4 Measurement of Adhesion

Four strips 2 cm wide were scored on each sample. Each strip in turn was peeled at 180° on an Instron tensile test machine at a cross-head speed of 50 mm min^{-1}. The peel load was taken as the average load per unit width during peeling.

3. RESULTS OF PEELING

3.1 Unbacked samples

Figure 1 shows the peel loads found for peeling unbacked strips of each of the four polymers from the substrates studied. Even from a superficial glance it is clear that both polymer and substrate have an effect on the adhesion.

The results in Fig.1 refer to a standard thickness of polymer. The thickness of the strip has a marked effect on the measured peel load, as shown in the results for polymer A in Fig.2.

A conspicuous aspect of peeling the unbacked polymer is the considerable tensile deformation of the freed strip, sometimes up to 300%. Coatings were also prepared and tested with a practically inextensible fabric backing to the polymer in order to suppress this phenomenon.

3.2 Backed samples

Results of peel load for the backed samples are shown in Fig.3, which should be compared with the similar presentation in Fig.1 for the unbacked polymer. There is a general similarity between the two figures with respect both to the effect of substrate and that of polymer. However the magnitude of adhesion is generally higher for the backed polymer, especially for the chlorite-formed films on copper. Thus further consideration was given to the three factors which these results imply are significant: the mechanical properties of the polymers,

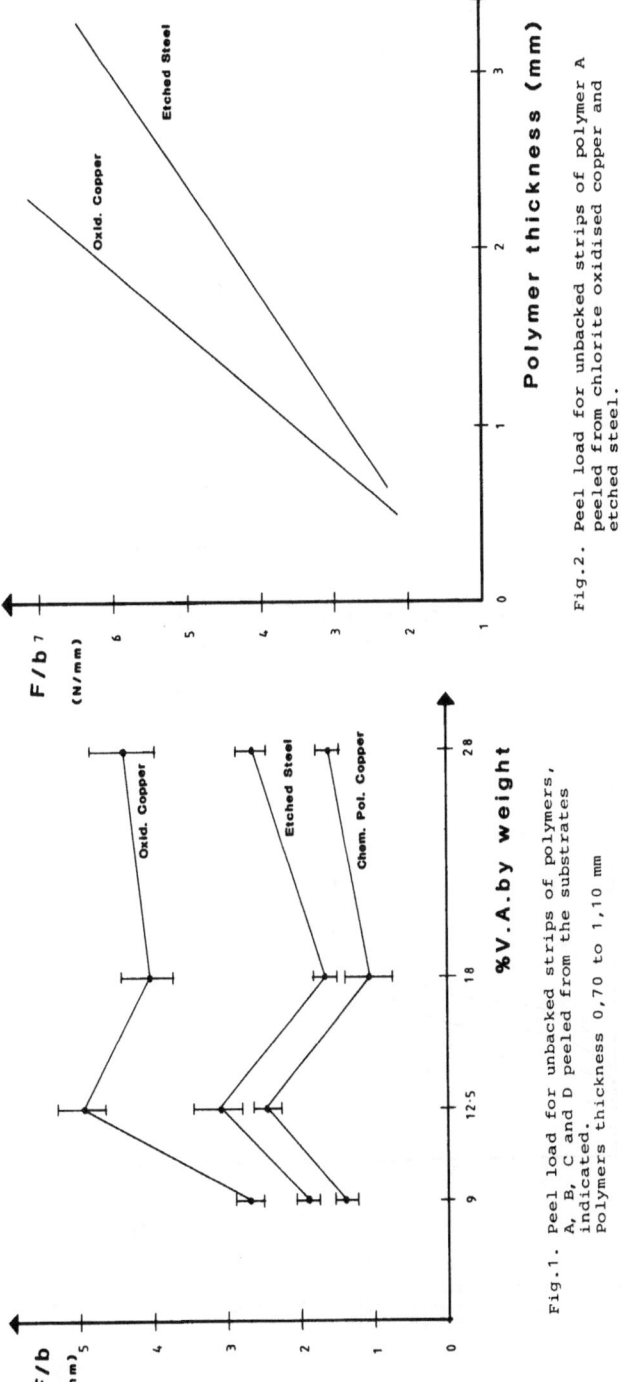

Fig.1. Peel load for unbacked strips of polymers,
A, B, C and D peeled from the substrates
indicated.
Polymers thickness 0,70 to 1,10 mm

Fig.2. Peel load for unbacked strips of polymer A
peeled from chlorite oxidised copper and
etched steel.

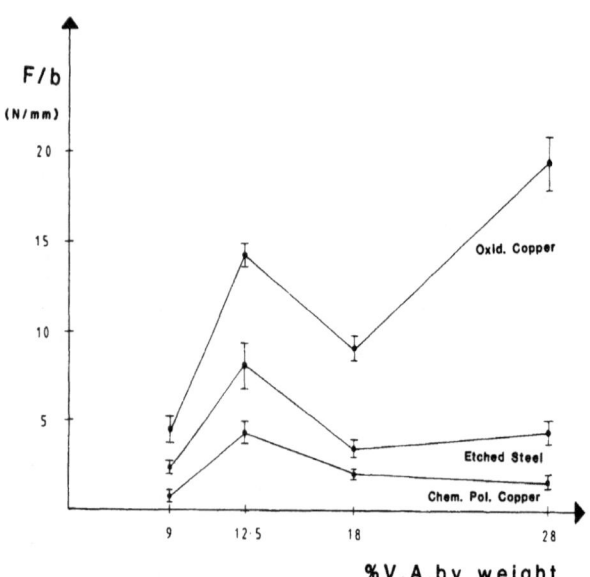

Fig.3. Peel load for backed strips of polymers A, B, C and
D peeled from the substrates indicated.
Polymer thickness 1,50 to 1,70 mm.

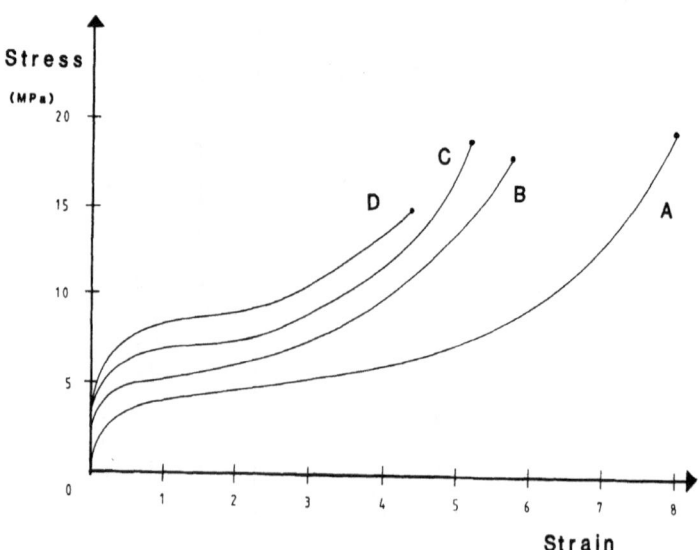

Fig.4. Stress-strain curves for polymers A, B, C and D
tested at a strain rate of 1,38 min⁻¹.
Dumb-bell shaped specimens were used.

the nature of failure at the interface and the effect of backing on the mechanics of peeling.

4. MECHANICAL PROPERTIES

The tensile properties of the polymers were studied using standard dumb bell shaped specimens. The curves obtained are shown in Fig. 4. The trends shown are broadly as expected[4]. The strain energy densities of the polymers at failure were evaluated and are given in Fig. 5. Standard tensile properties are summarised in Table 2.

Table 2

Tensile properties (with 95% confidence limits) of polymers A, B C and D obtained from tests at a strain rate of 1.4 min^{-1}.

	D	C	B	A
Young's modulus (MPa)	110.6	86.9	57.3	16.6
	±2.0	±2.4	±5.5	±1.1
Tensile strength (MPa)	14.9	17.1	16.6	19.0
	±0.5	±0.3	±0.6	±1.2
Elongation at break (%)	421	538	575	843
	±71	±15	±22	±37

As an alternative measure of fracture energy tear tests were performed on "trouser" type specimens the legs of which carried fabric backing to prevent the high extension of the torn strip. The results are given in Fig. 6.

5. THE FRACTURE SURFACES

5.1 Scanning electron microscopy

Observation of both sides of the fracture surface enabled information to be obtained about the locus of the failure and a qualitative assessment to be made of the extent of interfacial deforamtion.

The chlorite formed films on copper lead to cohesive failure in the polymer with extensive surface deformation usually drawing the polymer into filaments (Fig. 7a). The steel substrate also showed cohesive failure, but with less

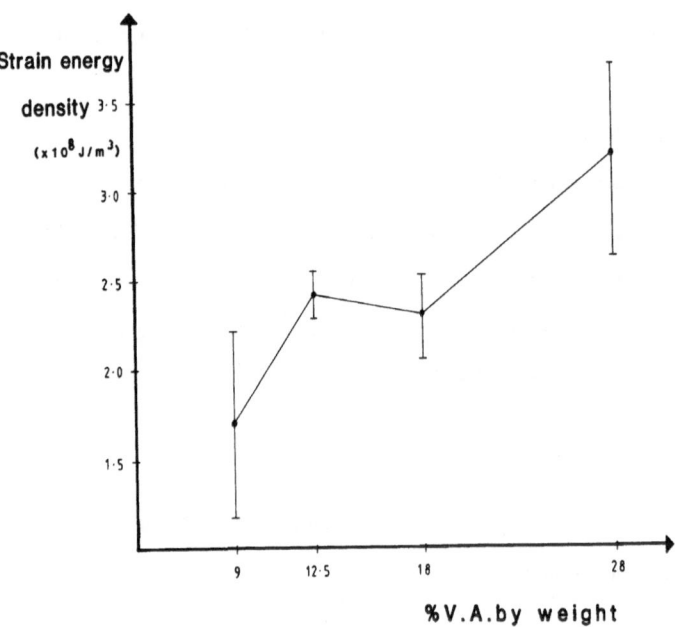

Fig.5. Strain energy density at failure of polymers A, B, C and D obtained from tensile tests.

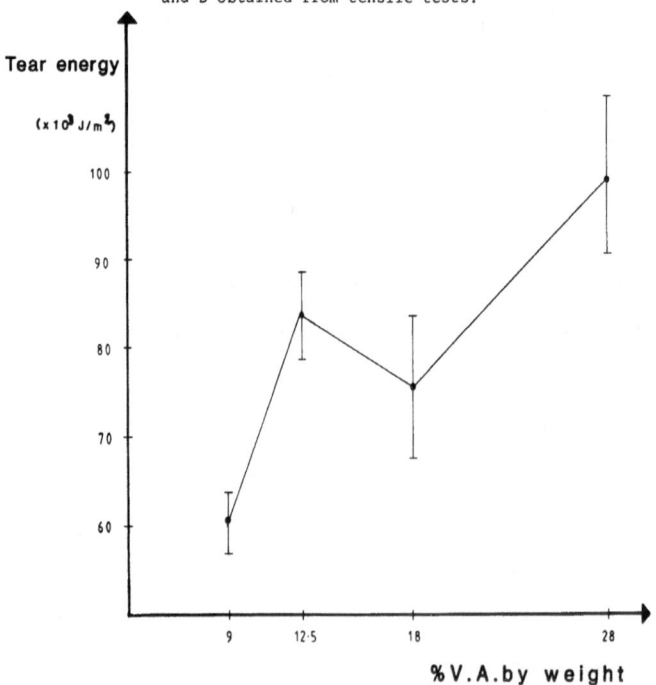

Fig.6. Tear strength of polymers A, B, C and D expresssed as force per unit thickness to propagate a tear in a "trousers" type specimen.

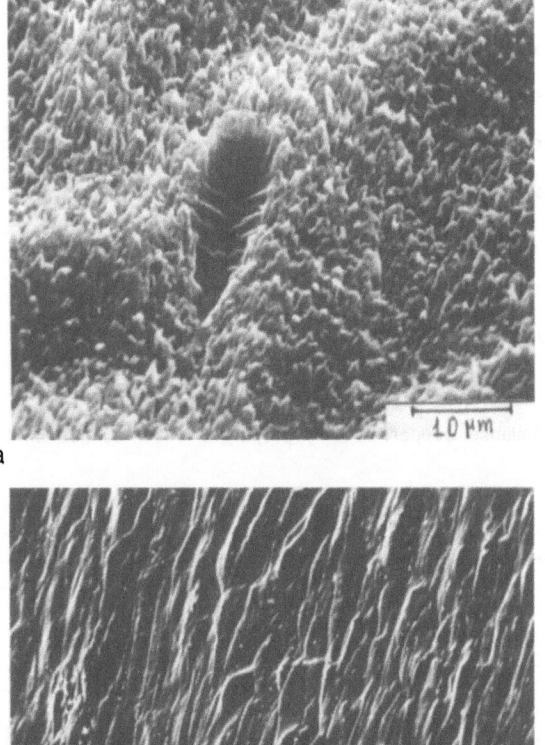

a

b

Fig.7. SEM picture of polymer side of fracture
of a backed strip of polymer A peeled from
(a) chlorite oxidised copper and (b) polished
copper.

obvious drawing of the polymer. The polished copper gave surfaces with still fewer signs of plastic deformation of the polymer in the surface (Fig. 7b). The SEMs of this substrate suggest that a thin layer of residual polymer may be present, but they are not unambiguous.

Comparison of contact angles measured (a) on the peeled polymer strip and (b) the chemically polished copper from which the strip had been peeled with those on (c) the same substrate to which no polymer had been applied (Table 3) support the idea of cohesive failure in this case.

Table 3

Contact angles (degrees) of three liquids on the surfaces of

(a) backed polymer B peeled from chemically polished copper,

(b) the copper surface from which this polymer had been peeled,

(c) chemically polished copper, freshly prepared. (95% confidence limits are indicated).

	Polymer B peeled surface	Coppersubstrate after peeling	Fresh chemically polished copper.
Water	59.3±0.9	59.4±1.6	32.1±1.4
Formamide	59.6±1.4	61.0±0.8	48.1±1.8
Glycerol	84.4±2.2	84.4±1.8	73.1±1.0

Thus while the nature of the SEMs was broadly characteristic of the substrate involved there were only relatively small distinctions comparing polymer with polymer for the same substrate. There is, for example, no dramatic difference in appearance of the backed strip peeled from chlorite-oxidised copper between polymer A (peel load 20 N/mm) and for polymer D (4 N/mm), despite the difference in adhesion.

5.2 Thickness of Residual Polymer

In similar experimental circumstances to those being discussed the thickness of residual polyethylene has been found to increase with peel strength[2]. The residual polymer after peeling backed strips of the four EVAs from chlorite-oxidised copper was freed form the substrate by treatment with dilute hydrochloric acid. The freed film was floated onto a piece of cellophane film and, after drying, its thickness was directly measured in the SEM by observing the profile at almost zero angle. Obviously the "thickness" of such a heavily drawn film can have no absolute value, but it is considered that the relative values are comparable. The results are given in Table 4 together with the peel load corresponding to the particular film. For a given polymer the range of peel load is not sufficiently wide to enable the correlation with residual polymer thickness to be tested. It is clear, however, that from polymer to polymer there is no positive correlation between residual thickness and peel load.

Table 4

Adhesion of backed EVAs to chlorite formed films on copper. Residual thickness (with 95% confidence limits) and peel load for polymers, A, B, C and D.

A	Thickness (μm)	1.1±0.1	1.3±0.2	1.4±0.2
	Peel load (Nmm^{-1})	15.4	20.6	23.0
B	Thickness (μm)	2.5±0.5	2.2±0.2	2.7±0.5
	Peel load (Nmm^{-1})	7.8	8.5	10.2
C	Thickness (μm)	1.1±0.1	2.1±0.3	1.3±0.2
	Peel load (Nmm^{-1})	13.2	14.0	14.4
D	Thickness (μm)	2.6±0.3	2.5±0.2	2.3±0.2
	Peel load (Nmm^{-1})	3.8	4.9	5.0

6. DISCUSSION

6.1 General

For each category presented in Figure 1 and Figure 3 the peel load increases with change of *substrate* in the order chemically polished copper, etched steel, chlorite oxidised copper. This is reminiscent of results for the adhesion of low density polyethylene, detailed explanations of which can be found elsewhere[2,5,6]. In brief, the steel promotes chemical changes in the polymer favouring adhesion; the copper inhibits these changes. However, the microfibrous surface of the chlorite-oxidised copper encourages plastic deformation in the bulk polyethylene.

The SEMs of the EVA fracture surfaces, discussed above, show a similar progression in surface deformation to that found with polyethylenes. In comparing the peel loads of the different EVA *polymers* (Figs. 1 and 3) it is useful to look for similar trends in their mechanical properties. Unlike the peel loads, some properties, such as modulus, yield strength and elongation at break change monotonically with vinyl acetate content along the series A, B, C, D. Other properties, in particular the strain energy density at failure (Fig. 5) and tear energy (Fig. 6), show a trend reminiscent of the peel loads': values for D tend to be low, for A high and there is a peak at C. The enhanced properties of C for its vinyl acetate content are presumably associated with a somewhat higher molecular weight[4] implied by a lower melt index (Table 1). Thus on a superficial level it appears that the peel loads of the EVAs increase as the toughness-energy to break-of the polymer increases. This does not necessarily correlate with vinyl acetate content because difference in molecular weight are a complicating factor.

6.2 Energy Dissipation

The work reported in this paper is continuing. It is possible however to speculate usefully about the results in more detail.

It must be remembered that the peel load per unit width of strip is equivalent to an energy of fracture per unit area. Thus it represents all work done during peeling under the particular experimental conditions employed. The work X done by the test machine may be expressed as

$$X = W + \Psi \qquad \qquad \qquad (1)$$

where W is the Dupré work of adhesion or of cohesion (according to the failure mode) and Ψ represents all other types of energy dissipation. The polarity of the EVAs increases with vinyl acetate content so the value of W would increase in the order D to A. The magnitude^2 of the thermodynamic term however is tiny compared with that of X. Thus our attention is concentrated for the time being on Ψ.

When the polymer is unbacked, the freed strip stretches considerably (v.3.1 supra) absorbing energy Ψ_s, thus

$$X = W + \Psi' + \Psi_s \qquad \qquad \qquad (2)$$

where

$$\Psi' = \Psi - \Psi_s \qquad \qquad \qquad (3)$$

By equating the work done by the machine to the sum of the work done in stretching the freed strip and the remainder of the peel energy $(W + \Psi')$, it is easily shown for 180° peel that

$$W + \Psi' = F/b(1+\lambda) - Et \qquad \qquad (4)$$

where

F is the peel force,

b is the width of the strip,

t is the thickness of the strip,

λ is the extension ratio to which the freed strip is stretched.

E is the strain energy density of the strip at elongation ratio λ

and the term $F/b(1+\lambda)$ corresponds to X.

This equation has been applied by other workers[7-10] to elastically deformed strips, but will apply, mutatis mutandis, beyond the yield point.

This goes some way to explaining the thickness dependence of peel load shown in Fig. 2. The thicker strips require more energy for deformation. The energy term $W + \Psi'$ (Equation 4) was calculated for these results and is plotted against thickness in Figs 8 and 9. There is still a marked thickness dependence of energy, but it is less.

In peeling, the polymer strip is bent back, thus there is the possibility of viscoelastic and, if the yield stress is exceeded, plastic energy losses in both backed and unbacked strips. Equation 2 can be rewritten making the bending losses explicit:

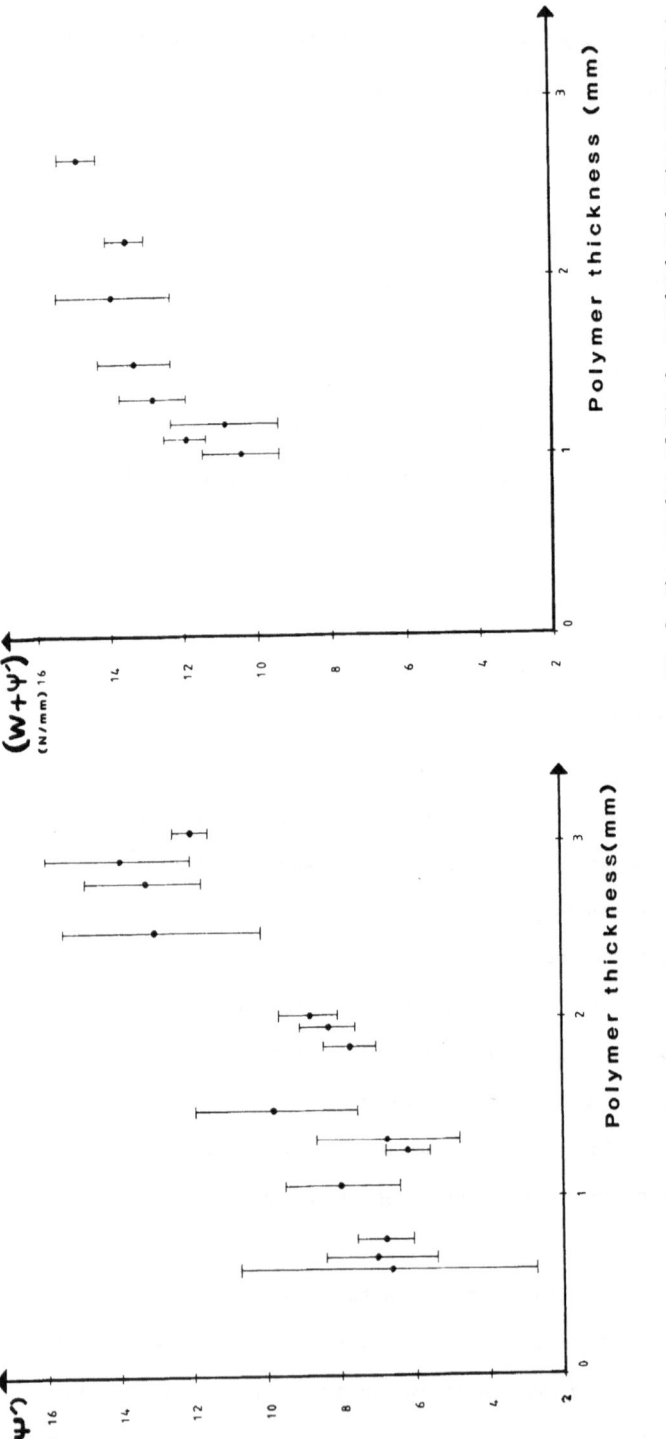

Fig.8. The results of Fig.2 recalculated using Equation 4
to give the variation of the energy W + ψ'
with polymer thickness for unbacked strips of
polymer A peeled from etched steel.

Fig.9. The results of Fig.2. recalculated using Equation 4
to give the variation of the energy W + ψ' with
polymer thickness for unbacked strips of polymer
A peeled from chlorite oxidised copper.

$$X = W + \Psi_0 + \Psi_B + \Psi_s \qquad . \quad . \quad . \quad . \quad . \quad . \quad . \quad (5)$$

where Ψ_B is the energy lost in bending and

$$\Psi_0 = \Psi' - \Psi_B \qquad . \quad . \quad . \quad . \quad . \quad . \quad . \quad . \quad (6)$$

The problem of evaluating Ψ_B has been considered by Gent and Hamed[11] and Crocombe and Adams[12]. Gent and Hamed treat their polymer (polyethylene terephthalate) as an ideal elastic-plastic body. Simple theory of plastic bending then gives:

$$\Psi_B = 0.5 \, \sigma_y e_y t \, [t/2Re_y + 2Re_y/t - 2] \, . \quad . \quad . \quad . \quad . \quad (7)$$

where

σ_y is the yield stress.

e_y is the yield strain.

t is the thickness of the strip.

R is the minimum radius of curvature of the neutral section.

With this in mind, the stress strain-curves (Fig. 4) of the EVAs were idealised to conform with an elastic-plastic model, and some values of R were measured from photographs of the peel front. Equation 7 was then used to get a rough idea of Ψ_B for comparison with the peel results in Fig. 3. Results so far available suggest that for polymer D (with the highest yield strength) the bending losses may be responsible for around 25% of the peel loads in Fig. 3. The figures for C, B and A are about 15-20%, 10% and 4-5% respectively.

The situation for the unbacked samples is slightly different as these are strained in tension at the same time as being bent. If it may be assumed that the strip had already yielded throughout its thickness owing to the tensile force and that the tension and bending can be treated as additive then for an elastic plastic body treatment analogous to Gent and Hamed's [11] gives

$$\Psi_B = \sigma_y t^2/2R \, . \quad . \quad . \quad . \quad . \quad . \quad . \quad . \quad (8)$$

The minimum possile value of R is $t/2$ which occurs when the strip bends right back on itself, so the maximum value of Ψ_B is $\sigma_y t$.

Considering the results of Figs. 8 and 9 it might be thought appropriate to substract $\sigma_y t$ from the higher values of peel energy, and perhaps something less from the lower values as the radius there would be greater. A rough calculation suggests that this would remove much of the thickness-dependence from the results for steel, but, perhaps significantly, less of the dependence from those for chlorite oxidised copper.

Thus in Equation 5, which may be written as

$$X = F/b \; (1+\lambda) = W + \Psi_0 + \Psi_B + \Psi_s \; . \qquad . \qquad . \qquad . \qquad (9)$$

the likely magnitude of the bending and stretching energy term, Ψ_B and Ψ_s, has been considered and their contribution to the measured peel load, F/b, assessed. It is clear that the term Ψ_0, which is a sort of "residual peel energy", is important, especially where the peel load is particularly high in Fig. 3.

The term Ψ_0 includes all the energy dissipation processes at and in the region of the peel front. The energy associated with surface drawing such as is shown in Fig. 7a and plastic and viscoelastic losses throughout the thickness of the polymer occurring as this surface forms will all form a part of Ψ_0. While the absolute value of W is small compared with X (or Ψ_0), its magnitude is relevant as it influences the magnitude of Ψ_0. In cases where the equivalent of Ψ_0 was chiefly determined by viscoelastic losses Andrews and Kinloch[13] and Maugis and Barquins[9,10] have found that the term corresponding to Ψ_0 is proportional to W. For higher values of W, higher stresses can be transmitted to the bulk polymer so greater losses occur.

In Fig. 3 the higher peel loads are associated with polymer C (which has a high intrinsic toughness), with polymer A (which has a high surface energy) and with chlorite oxidised copper (which would be expected to have a high surface energy). These factors appear to combine to give highest peel loads for A and C with the chlorite oxidised copper.

One way in which the losses included in Ψ_0 can be enhanced can be seen by careful observation of the geometry of the peel front for backed specimens. In many cases the peeling strip appears to curve tangentially to the substrante (Fig. 10a). Sometimes however a distinct "lip" of very highly extended

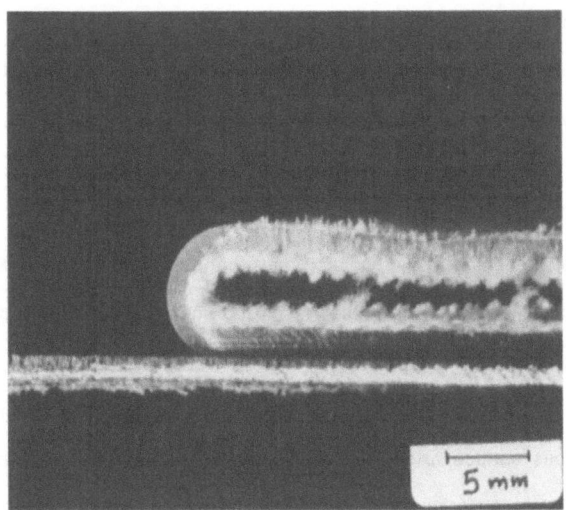

a

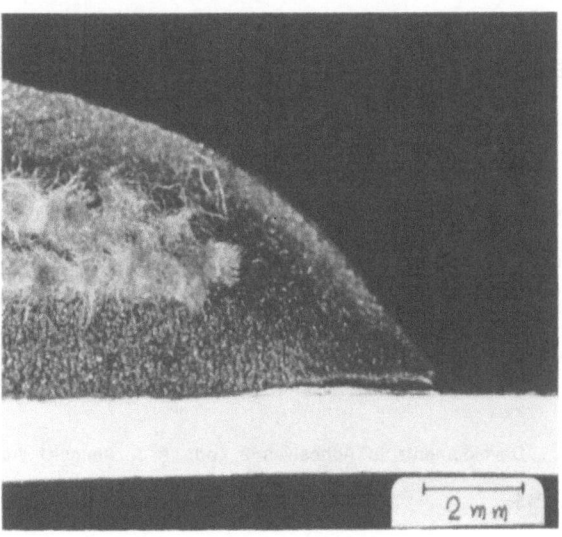

b

Fig.10   Picture of the peel front profile of
(a) polymer D and (b) A peeling from
chlorite oxidised copper.

polymer such as was observed by Satas and Egan[14] in a different context, remains on the substrate a short way behind the peel front (Fig. 10b). The tendency to form such a lip is greatest where the yield strength is lowest (i.e. polymer A) and for the chlorite oxidised copper. This combination of course gives the highest peel load (Fig. 3).

6.3 Summary

The results presented here suggest that the adhesion of these EVAs to metals reflects the mechanical properties of the polymer and the nature of the substrate. High toughness in the polymer and also low yield strength can enhance adhesion. Of the substrates studied the highest adhesion was found to copper with a fibrous surface oxide film into which the polymer can penetrate.

In the peel test stretching of the freed strip and energy losses in bending contribute to the practical adhesion but the major contribution usually comes from losses around the peel front itself.

7. ACKNOWLEDGEMENTS

Provision of a research studentship (for T.A.H.) by the University of Bath Research Fund and of electron optical facilities in this School by the SERC are gratefully acknowledged. We wish to thank the Materials Quality Assurance Directorate (Ministry of Defence) for providing the cotton duck used.

8. REFERENCES

1. Wake, W.C. Adhesion and the Formulation of Adhesives, 2nd Edn. Applied Science, 1982.

2. Packham, D.E., Developments in Adhesives-2 (ed. A.J. Kinloch) Applied Science, 1981, p.315.

3. Evans, J.R.G. and Packham, D.E. J.Adhesion 10, 39, (1979).

4. Gilby, G.W., Developments inRubber Technology-3, (ed.A. Whelan and K.S. Lee), Applied Science, 1982, p.101.

5. Bright, K. and Malpass, B.W., Europ. Polym.J., 4, 431 (1968).

6. Evans,J.R.G. and Packham, D.E., J. Adhesion, 10, 177 (1979).

7. Lindley, P.B., J.I.R.I., 5, 243 (1971).

8. Kendall, K. J. Phys. D. (Appl. Phys.), 8, 1449 (1975).

9. Barquins, M., Thesis for D. ès S., Université P. et M. Curie, Paris, 1980, p. 150.

10. Maugis, D. and Barquins, M., Adhesion and Adsorption of Polymers (ed. L-H. Lee), Plenum 1980, p. 203.

11. Gent, A.N. and Hamed, G.R., J. Appl. Polym. Sci., 21, 2817, (1977).

12. Crocombe, A.D. and Adams, R.D., J. Adhesion, 13, 241 (1982).

13. Andrews, E.H. and Kinloch, A.J. Proc. Roy. Soc. A322, 285(1973).

14. Satas, D. and Egan.F., Adhesives Age, 9(8), 22(1966).

Chapter 4

Formation of indium bonds for ultrasonic systems   and
examination of metal diffusion bonds by scanning acoustic
microscopy

H. S. Frew, B. S. Johal   and   M. Nikoonahad

Department of Electronic and Electrical Engineering
University College London
Torrington Place
London WC1E 7JE
England

ABSTRACT

Efficient transmission of ultrasonic radiation between two solids relies on
the properties of the bond or adhesive layer between them.
Thermocompression indium bonding in vacuum is a well known technique for
producing acoustically transparent bonds. We have developed a new bonding
technique with indium at atmospheric pressure. Measurements on the
acoustic and mechanical characteristics of such bonds are presented. It is
shown that when bonding a transducer, at frequencies in the range of 100
MHz, indium bonds have considerable advantages over organic adhesive
bonds.

A new form of microscopy (scanning acoustic microscopy) based upon imaging
with ultrasonic waves has come into existence in recent years. One of the

major applications of the acoustic microscope is nondestructive imaging through optically opaque solids, with high spatial resolution. The operation principle of the acoustic microscope is outlined and example micrographs, obtained from a range of metallic bonds, are presented.

1 INTRODUCTION

In ultrasonic systems there are often situations where one wishes to bond two solids together. An example is when one wishes to transfer acoustic energy from one solid sample to another for whatever reason. The quality of the bond in such situations plays an important role in the transmission characteristics of the bond. It can easily be shown that in order to have a transparent bond, between two identical solids, the acoustic impedance (that is the product of mass density and sound velocity) of the bonding agent must be as close as possible to that of the two solids [1]. It can also be shown that if the impedance of the bonding agent differs greatly from that of the samples, one needs to resort to a very thin bonding layer. Indium offers two advantages: Firstly, its acoustic impedance is comparable to that of many commonly used solids. Secondly it can be obtained in the form of a thin film by vacuum deposition techniques.

Our particular interest in this is to fabricate efficient transducers for scanning acoustic microscope (SAM) systems,[2,3] which also provide a theme for this paper. The basic elements of a SAM operating in reflection mode are shown in fig 1. The specimen to be imaged is immersed in a fluid coupler (usually water) which acts as a coupling medium. The ultrasonic field generated by the transducer is brought to a focus on the object and the reflected acoustic energy from the object is collected and recollimated by the lens and detected by the transducer. The voltage produced by the transducer upon reflection is used to intensity modulate an electron beam on a TV monitor and the lens is subsequently scanned over the entire specimen and thus the image is constructed point by point. There are two motivations for using an acoustic microscope. Firstly: the contrast of the image depends on the variation in the mechanical properties of the specimen; that is the density, elasticity and viscosity. Secondly: the acoustic microscope can be used to image through optically opaque solids. These two make the acoustic microscope an attractive tool for a class of problems in non-destructive evaluation (NDE), [4].

We see that the heart of the SAM is the transducer. We are concerned with frequencies in the range of 1 MHz to 5 GHz. Above 150 MHz the thickness of the transducer is only a few microns and one uses vacuum deposition techniques for depositing the transducer directly on the back of the lens rod. Below 150 MHz one uses a plate transducer which has to be bonded to the back of the lens rod - and it is in this situation that we require indium bonding. Indium bonding in vacuum is an established technique for bonding transducers [5]. However, it involves a number of complexities for manipulating the samples in the vacuum chamber. In the technique that we have developed, the bonding is carried out at atmospheric pressure which reduces the complexity of system,[6].

Section 2 is devoted to indium bonding. Section 3 deals with the applications of the scanning acoustic microscope for subsurface imaging and some results obtained from metal:metal diffusion bonds are presented. In section 4 we present a summary and the conclusions.

2 INDIUM BONDING AT ATMOSPHERIC PRESSURE

In this section we firstly present the bonding procedure and then describe our techniques for characterising the bonds. The preliminary results that we have obtained for quartz samples are also presented.

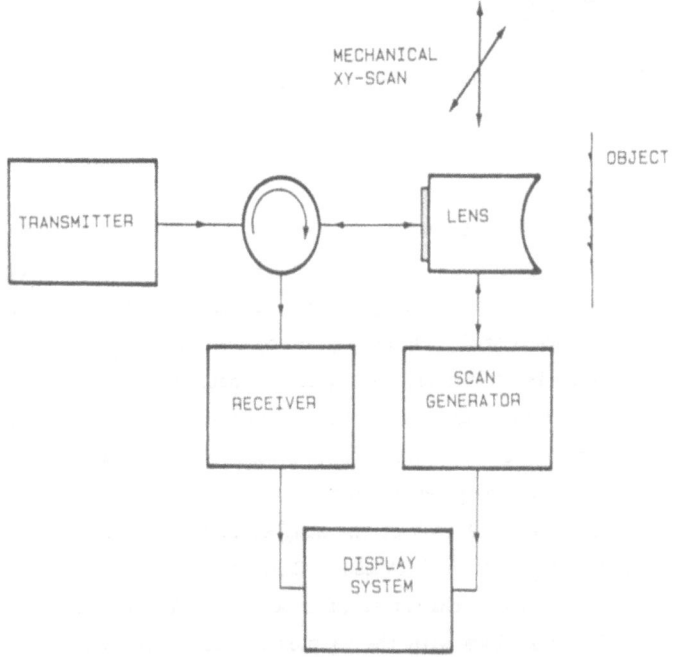

Figure 1.   The basic elements of a SAM operating in reflection mode.

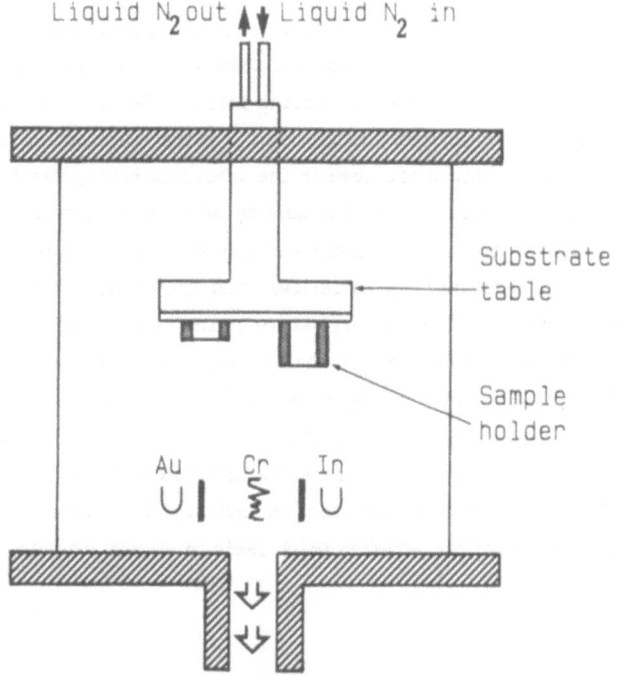

Figure 2.   Evaporater used to deposit CrAuIn films.

2.1 Formation of the Bonds

The samples which we bonded were quartz cylinders of 6mm diameter. Each
bond was made between a pair of cylinders, one of 3mm length, and one of
9mm length.

Following a standard cleaning procedure the samples to be bonded are placed
in a slightly modified evaporator where the samples can be cooled down to
liquid nitrogen temperature during evaporation, fig 2. In this evaporator
three metals can be evaporated - chrome, gold and indium and the typical
evaporation pressure is 10^{-6} torr. A liquid nitrogen trap is included
above the diffusion pump, to prevent hydrocarbon backstreaming, so that
this pressure is attainable. This independent liquid nitrogen trap remains
in operation throughout the evaporation process, which also prevents sample
contamination. One surface of each of the samples is first coated with a
flash of chrome (typically 200 Å thickness). Gold is then evaporated onto
the surfaces with a typical thickness of 1500 Å. Following the CrAu
evaporation, which takes place with the samples at room temperature, the
liquid nitrogen is introduced into the substrate table and the system is
allowed to reach an equilibrium. The thickness of evaporated indium films
is typically 1 μm.   We have found that it is absolutely essential that
the surfaces of the samples be cooled to a temperature significantly lower
than room temperature, before evaporating indium. Failure to meet this
requirment leads to nonuniform indium films. In the evaporator used, the
cooling provided by liquid nitrogen in the substrate table, was found to
make a considerable difference to the quality of films obtained. Figures
3(a)& (b) illustrate two SEM micrographs of indium films obtained with and
without liquid nitrogen cooling. We believe that the evident morphology in
the uncooled film is due to the very nature of the indium atoms. When the
uncooled indium atoms impinge upon the surface, they migrate (and hence
lose their energy) before coming to a complete adhesion. It is this
process which causes the nonuniformity of the films, [7]. From a
macroscopic point of view nonuniform films look "cloudy". It is believed
that when cooling the samples, the atoms on impinging lose their energy and
hence come to an immediate adhesion which leads to smooth and shiny films.

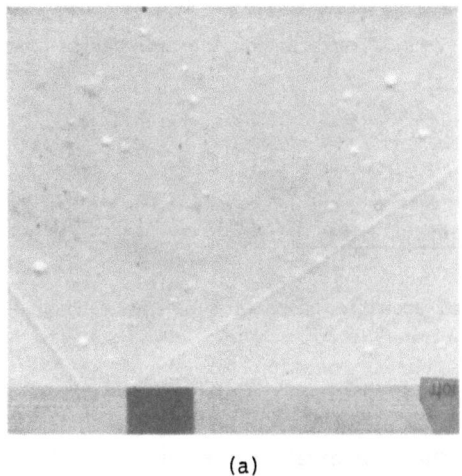

(a)

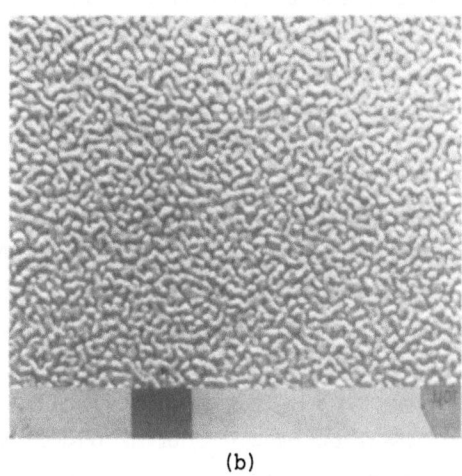

(b)

Figure 3.   SEM micrograph of an indium film.   (a) Without cooling,
(b) With cooling.

Following the indium evaporation, the samples were taken out of the indium evaporator and bonded at a temperature of 120°C. Typical bonding pressures ranged from 2 MN/m^2 to 12 MN/m^2, and the bonding time varied from 1 to 3 hours.

2.2 Characterisation of Indium Bonds

Both the mechanical and acoustic properties of the bonds have been characterised.

2.2.1 Mechanical test: The mechanical properties of the bonds were quantified by the stress required to pull the bonded specimen apart. One end of the sample was clamped to a fixed base plate, while the other was attached to the end of a wire, the other end of which was attached to a free hanging tray to which weights were added. This assembley prevents any shearing of the bond. The weight required to separate the samples serves as an indication of the adhesion of the bond.

2.2.2 Acoustic test: The acoustic quality of each bond was characterised by measuring its longitudinal wave reflection coefficient at 45 MHz. The basic set up for this measurement is illustrated in fig 4(a). Acoustic pulses (about 100 ns wide) are coupled to the bonded samples via a thin water layer. Figure 4(b) shows the reflections from various planes of the structure. Echo (A) is a spurious echo from the water bond; (B) is the echo from the indium bond. The better the quality of the bond the smaller the amplitude of echo (B). Echo (C) is due to that portion of the energy which is transmitted through the bond and is reflected from the back of the sample (which we assume is a 100% reflector). Echo (D) is due to a double transit in the short sample. A comparison between (B) and (C) easily leads to the value of the reflection coefficient.

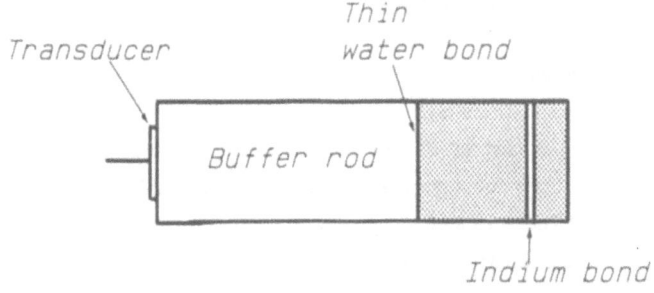

(a)

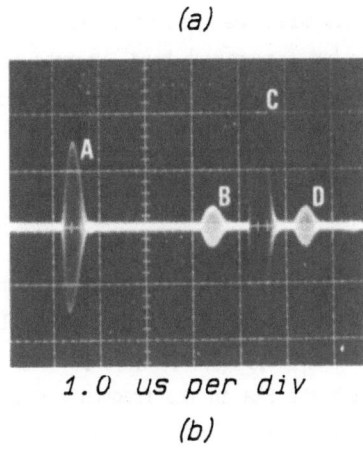

1.0 us per div

(b)

Figure 4.   (a) Experimental arrangement for measuring the
reflection coefficient of an indium film.   (b)
Reflections from the various planes of the structure.

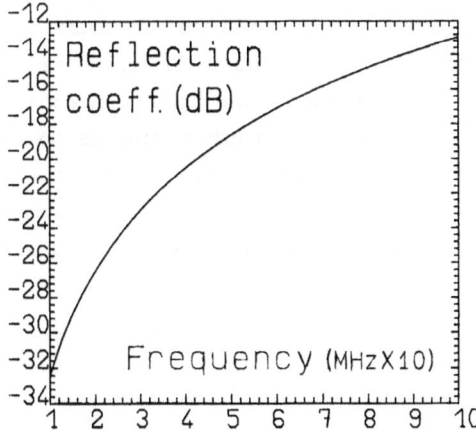

Figure 5.   Calculated variations of reflection coefficient with
frequency for a typical indium bond between two quartz
sections.   Cr, Au and In thicknesses.

2.3 Quartz:Quartz Indium Bonds - Results

The theoretical reflection coefficient for quartz:quartz bonds with typical thicknesses of Cr, Au and In is calculated and plotted as a function of frequency - fig 5. We see that at 40 MHz the reflection coefficient is -21 dB. Results obtained from four samples are presented in table 1. It is seen that in all cases there is a good agreement between the theoretical value of reflection coefficient and the measured ones. The results show no correlation between the mechanical breaking point of the samples and the acoustic properties. This might be due to the fact that a thin uniform film can lead to a fragile but transparent bond, whereas good bonding over a few patches can lead to mechanically strong but acoustically poor bonds.

2.4 Indium Bonds for Broadband Transducers

Generation of short ultrasonic pulses relies on the bandwidth of the transducer. It is not uncommon to encounter bandwidths larger than one octave in some NDE systems (typical frequency range is 1-20 MHz). Subsurface NDE imaging with the SAM in the frequency range of 50-150 MHz has many attractions and the incentive for using short pulses is to achieve the highest possible range resolution. For such frequencies, the bonding layer between the transducer and the substrate plays a major role in determining the transducer bandwidth. One way of achieving large bandwidth is to choose a bonding material which provides a good match between the transducer and the substrate. Figures 6(a)&(b) illustrate the calculated frequency characteristic of a 100 MHz lithium niobate transducer bonded to sapphire with an organic adhesive and with indium using the KLM equivalent circuit for the transducer [8,9]. We see that indium bonding in this situation gives rise to a bandwidth of well over 100% which means that one could generate a one cycle RF pulse leading to the highest possible range resolution at 100 MHz.

Sample	Applied pressure (MN/m²)	Temp. (°C)	Bonding time (hours)	Reflec. coeff. (dB)	Brk.ing stress (MN/m²)
1	2.1	120	3.0	-20	0.17
2	3.9	120	3.0	-15	0.07
3	7.8	120	1.5	-20	0.41
4	11.0	120	1.0	-17	>1.00

Table 1. Bonding conditions and results obtained for Quartz:Quartz bonds.

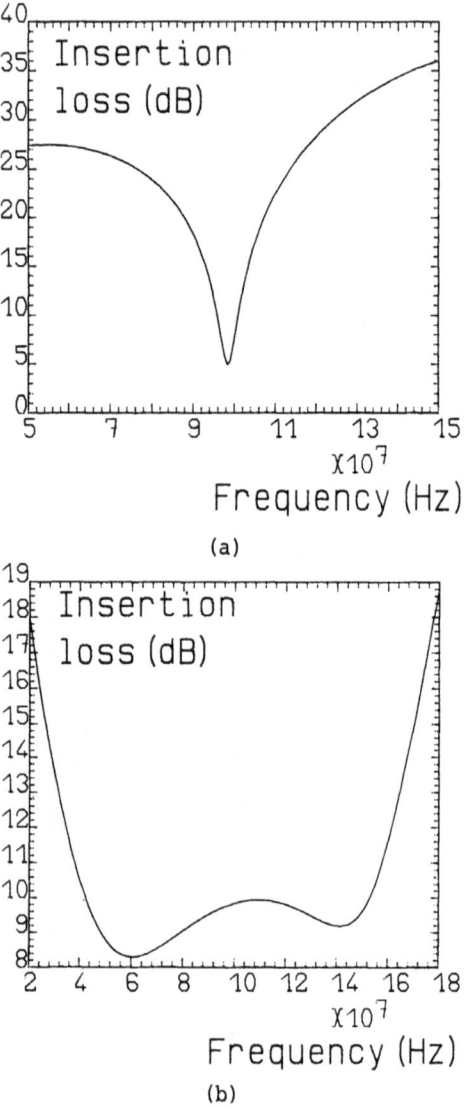

Figure 6.   The calculated frequency characteristic for a 100 MHz lithium niobate transducer bonded to sapphire (a) with an organic adhesive and (b) with indium.

3 APPLICATIONS OF THE SCANNING ACOUSTIC MICROSCOPE TO NDE

The basic operation of the SAM was outlined in the introduction. In material studies, applications of the SAM range from imaging microelectronic components to imaging the surface of metals and ceramics, [3,4]. As pointed out before a major application of the SAM is subsurface imaging through optically opaque solids. When imaging subsurface features, often an air gap (or indeed vacuum gap) of a few angstoms provides a large contrast. Figure 7 shows the reflection coefficient from an air gap between two half space steel layers. We see that the presence of an air gap of about 10 Å leads to almost 100% reflection, leading to a change of contrast by the same amount. SAM techniques therefore enjoy a fundamental superiority over other subsurface NDE methods (such as X-ray and neutron scattering techniques), for it is the mechanical nature of the object which is often to be tested. It is this capability which makes the SAM unique for adhesion tests. It has been shown that with suitably designed lenses one could achieve subsurface diffraction limited foci [10,11].This essentially means that a bonded sample can be examined with high spatial resolution (typically 100 μm resolution at 50 MHz). In the following we present a number of subsurface micrographs that have been obtained in the range of 40-60 MHz.

Figure 8(a) shows the schematic diagram of a power transistor package. The bonding between the Si chip and the copper header plays an important role in the thermal characteristics of the device. The acoustic beam was focussed through the header (which is a copper plate with 1 mm thickness) onto this bond, and the bond was imaged nondestructively. Figures 8(b) & (c) show two micrographs obtained from two different devices which illustrate the nonuniformity in the bonding. It is important to note that no sample preparation whatsoever is needed for such examination.

Metal:metal diffusion bonding is now a widespread technique for joining metals. One major application of this technique is in making aircraft parts. Various techniques of SAM provide a means for imaging such bonds, [12,13]. Figure 9 shows a bond between tungsten carbide and copper (the sample was a part of a cutting tool), imaged through 1.2 mm of copper.

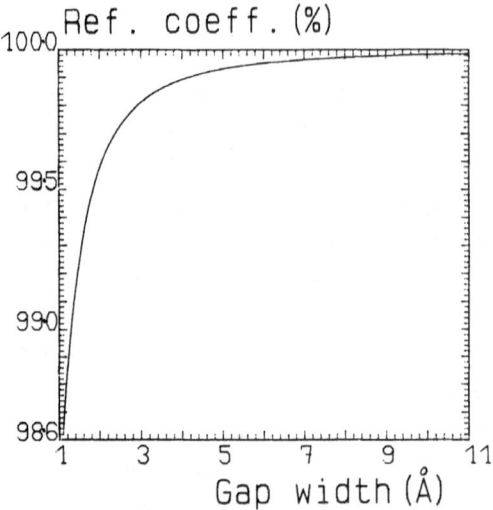

Figure 7. The calculated reflection coefficient of an air gap between two steel plates.

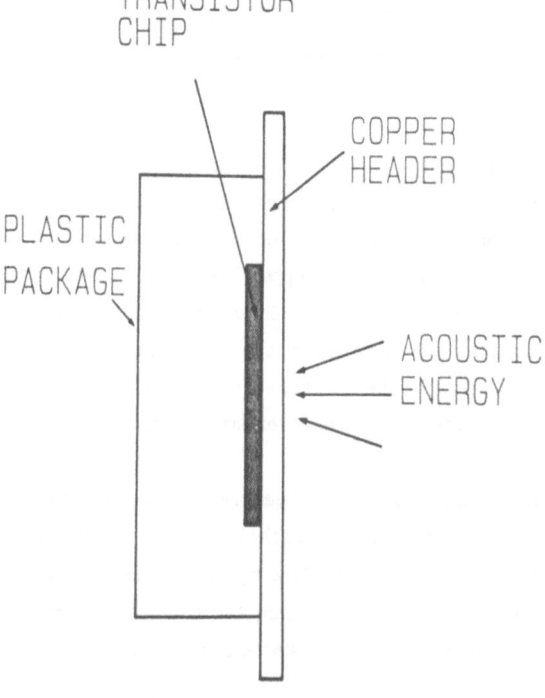

Figure 8. (a)

(b)

(c)

Figure 8.   (a) Schematic diagram of a power transistor package
            (b) & (c) Micrographs obtained at 40 MHz of two devices.
            Field of view: 6 x 6 mm.

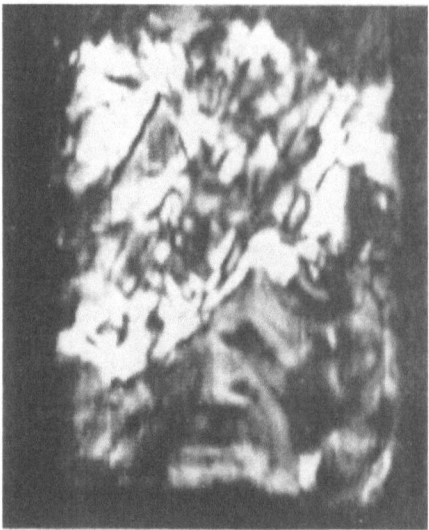

Figure 9.   Micrograph of a difussion bond between tungsten carbide
              and copper. Black regions indicate good bonding and
              white regions, poor bonding.

Black regions in the micrograph indicate regions of good bonding and white
regions illustrate poor bonding.

4 SUMMARY AND CONCLUSIONS

We have described a technique for bonding solid samples with indium and
presented some preliminary results that we have obtained from quartz
samples. Although the measured reflection coefficients are in good
agreement with the theoretical prediction, we believe that a number of
bonds have to be made before we can come to firm conclusions on optimum
bonding conditions. We have shown that lithium niobate transducers, indium
bonded to sapphire, exhibit a broad frequency response.

The operating principles of the scanning acoustic microscope have been
outlined and illustrated by some results which we have obtained for
metallic bonds.

ACKNOWLEDGEMENTS

We wish to express our gratitude to Professor E. A. Ash who kept a close
interest in this work. The work was supported by SERC, and in part by the
Wolfson Unit for Micro-NDE.

REFERENCES

[1]   Kinsler L. E. and Frey A. R., "Fundamentals of Acoustics", 2nd Edition, John Wiley & Sons Inc. New York, 1962.

[2]   Quate C. F., Atalar A. and Wickramasinghe H. K., " Acoustic Microscopy with Mechanical Scanning - a Review", Proc. IEEE 67(8),pp 1092-1114, 1977.

[3]   Nikoonahad M., "Recent Advances in High Resolution Acoustic Microscopy", Contemp. Phys. 25(2), pp 129-158, 1984.

[4]   Nikoonahad M., "Reflection Acoustic Microscopy for Industrial NDE" in "Research Techniques for Nondestructive Testing" (R.S. Sharpe Ed.), Vol 7, in press.

[5]   Sitting E. K., Warner A. W. and Cook H. D., "Bonded Piezoelectric Transducers for Frequencies beyond 100 MHz", Ultrasonics 7, 108-112, 1969.

[6]   Attal J., University of Montpellier - France, private communication.

[7]   Wyatt A. F. G., University of Exeter, private communication.

[8]   Krimholtz R. Leedom D. and Matthaei G., "New Equivalent Circuits for Elementary Piezoelectric Transducer", Elect. Letts 6, pp 388-389, 1970.

[9]   Desilets C.S., Fraser J.D. and Kino G. S., "The Design of Efficient Broadband Piezoelectric transducers", IEEE trans., SU 25(3), pp 115-125, 1978.

[10] Pino F., Sinclair D. A. and Ash E. A., "Scanning Acoustic Microscopy of Solid Objects using Aspheric Lenses", Proc. of 11th International Conference on Acoustical Imaging, Monterey.

[11] Nikoonahad M., Yue G.Q. and Ash E.A., "Subsurface Broadband Acoustic Microscopy of Solids using Reduced Aperture Lenses", in "Review of Progress in Quantitative NDE", (Thompson B.O. and Chimenti D.E. Ed.),Vol 2B, pp 1611-1623, 1982.

[12] Guangqi Y., Nikoonahad M. and Ash E. A., "Pulse Compression Subsurface Acoustic Microscopy", Elect. Letts. 18(18), pp 767-769, 1982.

[13] Yue G.Q., Nikoonahad M. and Ash E. A., "Subsurface Acoustic Microscopy using Pulse Compression Techniques", Proc. of IEEE Ultrasonics Symposium, pp 935 -938, 1982.

Chapter 5

ADHESION MECHANISM OF THICK FILM CONDUCTORS

M.V. COLEMAN

Standard Telecommunication Laboratories Limited
London Road, Harlow, Essex.

1 INTRODUCTION

Thick film conductors consist of a metallic phase which, after firing, adheres to 96% alumina substrates. The method of adhesion depends upon deliberate minority constituents added to the conductor ink which forms the bonding layer between the metal and ceramic[1]. The nature of this bonding layer and the adhesion mechanisms involved are critical in determining the necessary processing conditions[2]. The substrate type, its composition and structure, may also determine the effectiveness of the bonding layer.

Many of the conductor inks and substrate materials described were developed over a decade ago and some are no longer available or have been superseded by improved compositions.

2 FRITTED CONDUCTORS

2.1 The Effect of Firing Temperature

Thick film inks are dried at a temperature of typically 150°C for 20 minutes to drive off the organic solvents after printing. The ink pattern as defined by the printing process is maintained by an organic binder which holds the structure together. The patterned substrates are then fired through a conveyor belt furnace, experiencing a temperature profile. The firing time is typically 30 to 45 minutes with a rate of rise of between 50°C and 100°C per minute to a plateau at temperatures between 750°C and 1000°C, fig. 1.

A silver-palladium conductor, C4020, produced by Alloys Unlimited, printed on a Stemag substrate, was fired under various profiles giving peak temperatures between 600°C and 900°C. On each substrate there were five

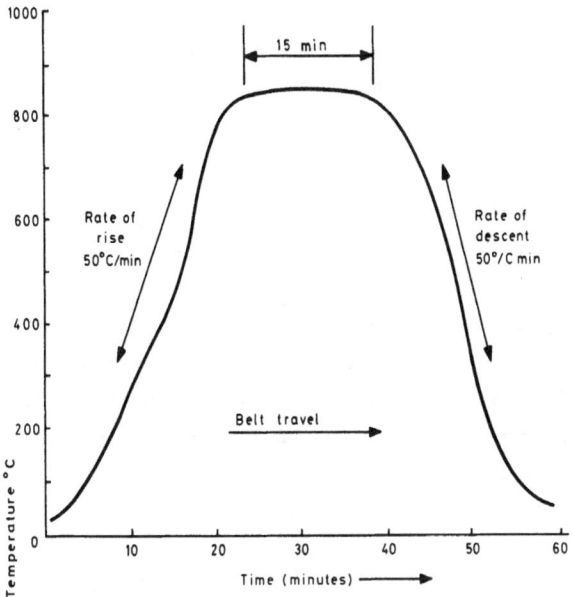

Fig.1   Typical conductor firing profile with a peak
temperature of 850°C

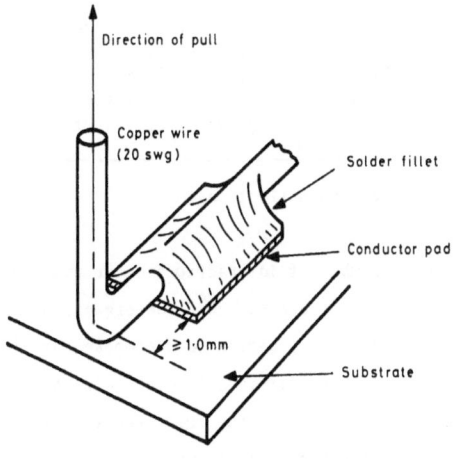

Fig.2. Diagram of pull - peel adhesion test

1.5 x 1.5 mm square pads on to which were soldered pretinned copper wires 0.7 mm in diameter. The wires were bent into right angles prior to attachment so that pulling vertically would produce a pull-peel test on the conductor pad, fig. 2. The solder used was 62/36/2 Sn/Pb/Ag with a mildly activated flux applied by dipping at a temperature of 230°C.

The adhesion strengths were measured on a Dage Precima micro tester and the values are shown in Table 1. At the low firing temperatures, the weakest part of the system was in the bulk conductor and failure was due to dissolution of the conductor in the solder. At the highest temperatures the conductor pads became detached from the substrate leaving a clearly defined square pattern in conductor glass. The highest adhesion strengths were found when the conductors were fired at 750°C and failures occurred due to pull-out of the wires from the solder in the conductor pads.

Table 1:
Average adhesion strengths (+2 standard deviations) of a
AgPd conductor

Peak Firing Temperature °C	Adhesion Strength kgf +2	Failure Modes
600	0.03*	leaching of conductor
650	0.23 + 0.27	leaching of conductor
700	0.42 + 0.16	glass-alumina interface
750	0.60 + 0.24	solder pull out
800	0.45 + 0.18	glass-metal interface
900	0.21 + 0.10	glass-metal interface

*All but one wire fell off at zero pull strength.

During the firing cycle, a number of interactions occur which may be described schematically as in fig. 3. At around 320°C the organic binder, which is usually based upon ethyl cellulose, begins to decompose, oxidise and burn off. By 450°C small amounts of flux additives interact with the conductor particles, aiding sintering and the glass frit starts to soften. At around 600°C sintering is occurring and the glass flows through the structure to the substrate.

At 750°C (for this particular conductor) the conductor glass has interacted with the minority constituents in the alumina substrate to form a

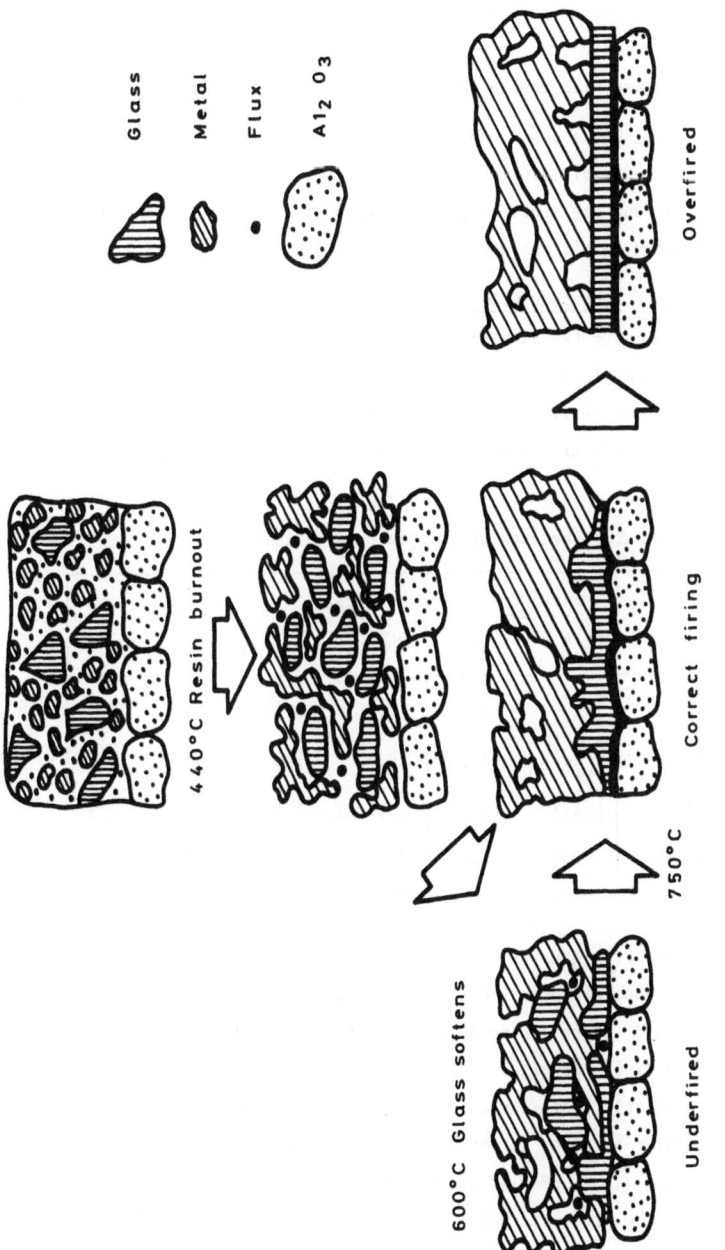

Fig. 3   Schematic of the effects of firing on conductor adhesion

good adhesive bond. The conductor has sintered and mechanical interlock is achieved between the conductor metal and the glass.

Firing at a higher temperature increases the level of sintering and permits further interaction of the glass with the substrate. The mechanical interlock of this glass with the conductor metal is reduced, however.

Firing at too low a peak temperature leads to inadequate sintering consistent with easy dissolution in solder and incomplete interaction between the conductor glass and the substrate. Firing too high reduces the conductor glass mechanical bond consistent with separation at that interface.

2.2   The Effect of Substrate Composition

Four conductors - two silver-palladium compositions, one gold palladium and one gold platinum - were printed on two manufacturers' substrates. Three peak firing temperatures were chosen and the adhesion strengths determined on 2 mm x 2 mm square pads by a pull-peel test, Table 2.

Table 2:
Adhesion strengths (kgf) on two manufacturers' substrates

| Conductor | 760°C | | 850°C | | 960°C | |
	A	B	A	B	A	B
AgPd C4020	2.0	2.1	2.8	2.4	2.6	2.4
AgPd DP 8151	1.4	1.0	0.7	1.5	1.1	1.1
AuPd DP 8227	1.7	2.1	3.1	2.3	3.3	2.3
AuPt DP 8641R	2.0	1.6	2.1	1.6	2.3	2.3

The adhesion strengths of both gold alloy conductors increased with firing temperature and, although the ink manufacturers' recommended temperature was 850°C, there was no evidence of the previously seen deterioration when fired too high. One of the silver-palladium conductors, C4020, behaved similarly. All three of these conductors exhibited significantly greater adhesion strengths on substrate A than B. The second silver-palladium conductor, DP 8151 did not follow this trend, with the lowest strength at 850°C, the preferred peak temperature, on Substrate A.

The composition of the two substrates was determined by Atomic Absorption Spectroscopy[3], Table 3. Both were debased alumina with calcium silicate as the main additional phase. The Smiths' substrate (A) contained more calcium silicate than the Stemag substrate (B). Visual examination of the second silver-palladium conductor on the Smiths' substrate revealed a yellow-green stain around the conductor about 1 to 2 mm beyond the nominal conductor edges.

Table 3:
Composition (weight %) of UK manufactured substrates

	A Smiths	B Stemag
Al_2O_3	95.4	97.3
SiO_2	2.2	1.7
CaO	1.9	0.8
MgO	0.1	0.1
Miscellaneous	0.4	0.1

The composition of this stained region was analysed on an SEM using x-ray analysis. The elemental intensity spectrum, fig. 4, showed that apart from aluminium, silicon and calcium that might be expected from the substrate, and lead and silicon from the conductor glass, there was a considerable amount of bismuth oxide. The bismuth oxide has interacted vigorously with the calcium silicate phase and this has not only led to bleed-out beyond the conductor pattern but a depletion of the bismuth phase in the conductor.

The role of the bismuth oxide is as a wetting agent or flux and aids both the wetting of the glass particles to each other and the glass to the substrate. There may also be some effect on the sintering of the metal powder. Dissolution in the substrate glass of the bismuth oxide reduces the effectiveness of these interactions, all of which are crucial to the overall adhesion strength of the conductor system. Whereas the UK manufacturers of substrates used a calcium silicate fluxed alumina ceramic, the Japanese and US manufacturers preferred a magnesium silicate debased alumina material, Table 4. The interaction of the bismuth oxide phase in the conductor with magnesium silicate is less vigorous so the conductor that exhibited bleed-out on the Smiths' substrates did not seem to suffer on the Kyoto, Coors or Alsimag substrates.

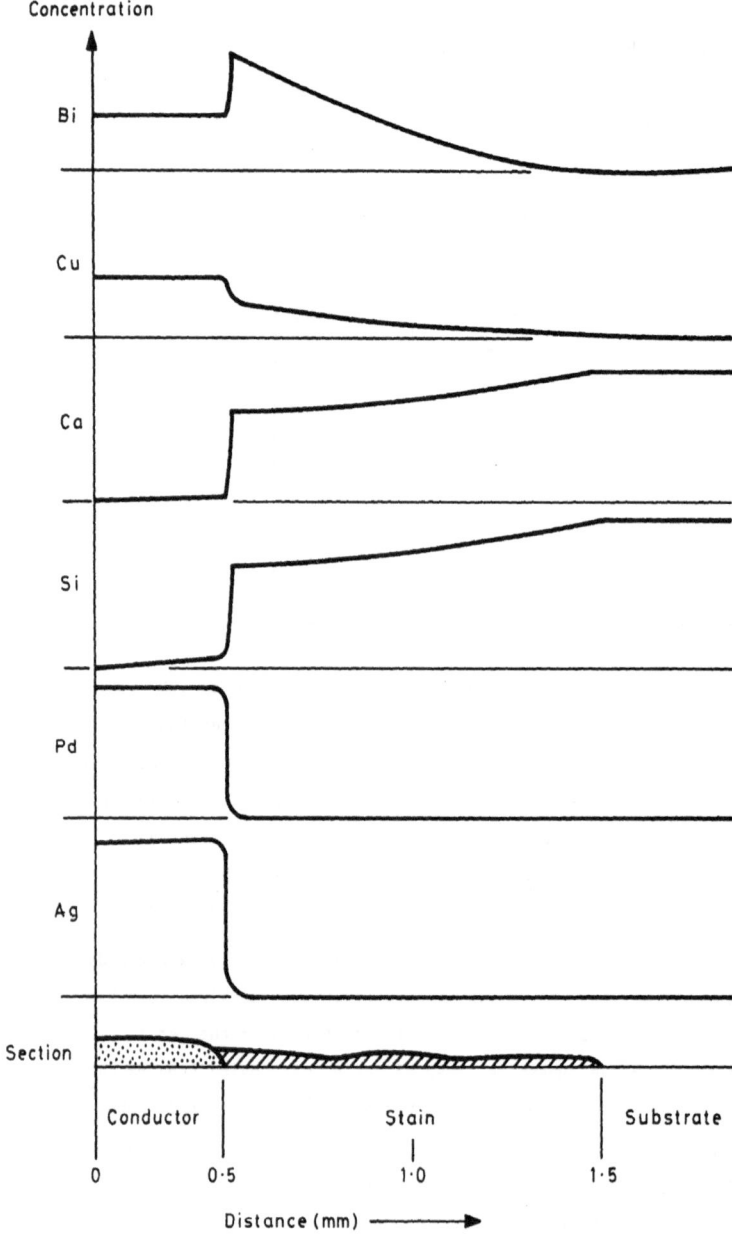

Fig. 4 Intensity spectrum of silver –palladium conductor
and adjacent stained region

Table 4:
Composition of 96% alumina substrates

	English Glass	Kyoto	Coors	Alsimag
Al_2O_3	97.0	95.3	95.1	94.7
SiO_2	2.0	3.4	2.3	3.0
CaO	1.0	0.1	0.0	0.2
MgO	0.1	1.4	1.8	2.2

2.3 The Effect of Substrate Structure

Alumina substrates are formed by sintering a slip of alumina and silicate particles at a temperature around $1600^{o}C$. The real surface area of the substrate is likely to be greater than the apparent geometrical area. The surface area depends upon the degree of sintering, the porosity and the depth of penetration of surface fissures into the substrate. This microstructure is too fine to be detected by surface profile techniques and so a gas adsorption technique was developed[4].

The surface area of one-inch square substrates was determined (total apparent geometrical area 13.6 cm^2), Table 5. The Alsimag and Stemag substrates had similar values of about 2.4 for the ratio of surface to geometrical area. However, Coors and Kyoto have much higher values. SEM micrographs of the surfaces of the four substrates showed that the grain size distribution was between 2 and 6 μm. Micrographs of the fractured surfaces showed that the Alsimag and Stemag substrates had dense structures. The Kyoto structure was similar but there was separation between the grains at the surface which gave rise to deeper gas penetration, and the Coors substrate showed fine fissuring between grains.

Table 5:
Surface areas of 96% alumina substrates

Substrate	Surface Area (cm^2)	Ratio of Surface Area To Geometrical Area
Kyoto	134.0	9.83
Alsimag	32.7	2.42
Coors	183.2	13.56
Stemag	32.9	2.43

Apparently, the silicate phase is more concentrated in the bulk of the substrate than the surface. The different grain structures at the surfaces may be related to different substrate processing, particularly firing, and this might affect the surface concentration of the silicate phase.

A silver-palladium conductor was printed and fired at a peak temperature of 850°C, (the ink manufacturers' recommended temperature) on to the Alsimag, Coors and Kyoto substrates. A 0.7 mm diameter copper wire was soldered to 3.8 mm x 2.5 mm pads with 60/40 Sn/Pb solder. The adhesion strengths were measured and the average strengths were 2.23 kgf, 2.18 kgf and 2.04 kgf respectively, with a standard deviation of approximately 0.4 kgf.

In all cases, the failures occurred due to partial pull-off of the solder from the conductor and there was clearly no significant difference in adhesion strengths. Therefore the surface structure, at least for the magnesium silicate containing substrates, does not seem to affect the adhesion strength. In comparison with the calcium silicate containing substrates, the adhesion strengths are higher, consistent with reduced interaction of the conductor glass with the magnesium silicate.

2.4   The Effect of Modified Conductor Composition

The conductor glass composition had been based upon non-crystallising glass with added bismuth oxide. A modified glass composition was produced by Du Pont which contained preformed bismuthate glass frit compositions. This glass when fired at 850°C partially devitrified giving barium-containing crystallites identified in a Du Pont patent[5] as $BaAl_2Si_2O_8$ (hexacelsian) which were visible as cubic crystals within the glass matrix.

The conductor glass that is left after the partial devitrification is
a higher softening point glass. Conductors based upon these glass
compositions are less sensitive to further firing operations than the previous
generation of conductor inks. The adhesion strengths were also found to be
higher, typically in excess of 2.5 kgf for 2 mm x 2 mm square pads. The
interaction with calcium silicate substrates was also less with no evidence of
bleed-out.

3   REACTIVELY BONDED CONDUCTORS

A generation of gold conductors was produced which were designed to
adhere directly to the alumina of the substrate rather than through the
silicate phase. The composition of conductors from various ink manufacturers
were determined, Table 6. All contained copper and some also contained
cadmium.

Table 6:
Composition of reactively bonded gold conductors

Ink Type	% Composition		Peak Firing Temperature ^{o}C
	Cu	Cd	
Cermalloy S4399	0.32	-	875 - 980
Du Pont 9500	0.83	0.18	950 - 1000
E-0 6990	0.70	0.34	970 - 1020
EMCA 3264	0.70	1.23	850 - 1020
Englehard T2888	0.62	0.50	980
ESL 8800	0.49	0.30	900+
Plessey C5700	0.67	-	1010 - 1035

It was noticeable that the peak firing temperatures recommended by the
ink manufacturers was much higher than for fritted conductors (the fritted
gold used as a comparison was Du Pont 9260 which had a recommended peak
temperature of 850°C). The reason for the high firing temperature is that
the reaction between the copper and alumina to form a spinel structure does
not occur until near the melting point of copper, 1083°C[6,7]. The addition
of the cadmium reduced the reaction temperature to below 1000°C and the

chosen peak firing temperature for the cadmium containing conductors was 975°C and for the others, 1025°C.

3.1   The Effect of Substrate Types

Three types of alumina substrate were chosen for comparing the adhesion strengths of the conductors, a) Alsimag 614, b) Coors 995 and c) Andermann and Ryder 974[8]. Copper wires were bent at right angles and soldered to 2.5 mm x 2.5 mm pads with a special solder 80:20 Au:Sn which inhibits leaching of the gold conductor. The adhesion strengths were determined, Table 7, and for comparison the values of the fritted conductor DP 9260 are included.

Table 7:
Adhesion strengths of golds on various substrates

Ink Type	Alsimag	Substrate Type Coors 995	A and R 974
Cermalloy S4399	2.45	2.60	1.50
Du Pont 9500	3.98	4.94	1.45
E-O 6990	3.96	3.77	1.34
EMCA 3264	2.03	1.83	1.42
Englehard T2888	3.35	3.70	2.79
ESL 8880	3.53	3.13	1.07
Plessey C5700	3.38	3.76	2.32
Du Pont 9260	1.26	0.78	1.11

The reactively bonded conductors had significantly greater adhesion strengths than the fritted conductor. Of the reactively bonded conductors, EMCA 3264 had much lower strength values on both Alsimag and Coors substrates. On visual examination it was evident that the ink had printed poorly such that the film was discontinuous.

As the interaction of the reactive phase within the conductor is with the alumina, the adhesion strength of any particular conductor should be substrate independent. The adhesion strengths of all conductors were considerably less, however, on the Andermann and Ryder 974 substrates than the

other two substrate types. In comparison the adhesion strength of the fritted
conductor was lowest on the high alumina substrate, Coors 995, and was only
slightly lower than the reactively bonded conductors on the Andermann and
Ryder substrate.

The composition of the three substrates was determined, Table 8. The
Alsimag substrate was a magnesium silicate debased alumina as previously
determined and the minority constituents in the Coors high alumina substrate
were a mixture of magnesium and calcium silicates. The Andermann and Ryder
substrate had been thought to be a calcium silicate based material but was
also a mixture of magnesium and calcium silicates in alumina.

Table 8:
Composition of alumina substrates

	Alsimag 614	Coors 995	A and R 974
Al_2O_3	94.7	99.2	96.9
AlO_2	3.0	0.5	1.4
CaO	0.2	0.1	0.8
MgO	2.2	0.2	1.0

The mode of failure on the Andermann and Ryder substrate was always
due to conductor pad removal from the substrate, whereas failures on the other
substrates were due to partial pad removal and wire pull-out from the solder
fillet. The surfaces of the Alsimag and Andermann and Ryder substrates were
examined on the SEM. The Alsimag substrate surface consisted of discrete,
well defined grains but the Andermann and Ryder substrate surface was glassy
in appearance with only a few of the grains defined. The surface topology of
the Andermann and Ryder substrate was determined using a Talysurf and the
trace showed that the surface was flat. The mode of manufacture of this
substrate involved producing a thicker substrate and then grinding the faces,
hence the flat surfaces.

3.2   The Effect of Lapped and Calcium Silicate Fluxed Substrates

The sensitivity to substrate of the adhesion strengths of the
reactively bonded golds was surprising. It was possible that either the
flatter surface or the presence of calcium silicate in the Andermann and Ryder
substrates contributed to the reduced adhesion strengths. Samples of

Alsimag 614 were lapped and the substrate surface was much flatter and the microstructure much finer after lapping compared with the Andermann and Ryder substrates, but not as smooth as the Coors 995 surface.

The adhesion strengths of three conductor inks were determined on lapped Alsimag, lapped Andermann and Ryder 93% alumina and a calcium silicate fluxed substrate, English Glass, Table 9.

Table 9:
Adhesion strengths on lapped and calcium silicate fluxed substrates

Ink Type	Lapped Alsimag	A and R 934	English Glass
E-O 6990	2.37	1.30	2.69
EMCA 3264	1.53	0.97	1.97
Englehard T2888	1.92	0.61	1.40

In all cases, the strengths were lower on the ground substrates and these had a higher proportion of the glassy phase at the top surface than the unground substrates. Examination of the areas on the ground substrates where the E-O 6990 gold had been pulled off, showed that the copper oxide was left behind and the underside of the removed pad was gold.

The copper (cadmium) aluminium oxide bond is formed with the alumina of the substrate, but the copper and cadmium also readily dissolved in the silicate phases of the substrate. The bond between the gold and the substrate relies not only on the interaction of the copper with the substrate but also on the formation of a copper gold bond, fig. 5. At temperatures in excess of $750^{\circ}C$, the copper alloys with the gold but at lower temperatures the copper comes out of solution and oxidises.

If there is excessive silicate at the substrate surface the copper and cadmium will dissolve in this phase and there will be a depletion in the gold near the substrate interface, fig. 6. This depletion will lead to reduced adhesion strength and failure at the gold substrate interface and explains the presence of copper on the substrate after pad removal.

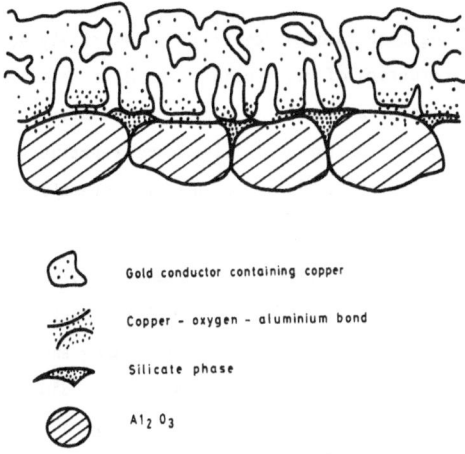

Gold conductor containing copper

Copper – oxygen – aluminium bond

Silicate phase

Al_2O_3

Fig.5 Schematic of reactive bond to alumina substrate

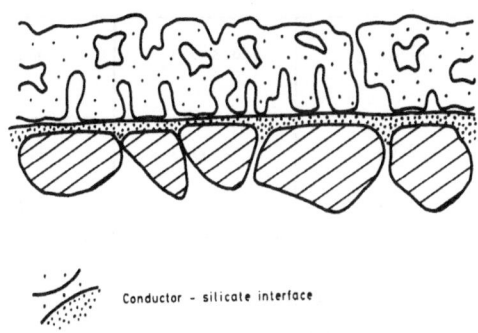

Conductor – silicate interface

Fig.6 Schematic of conductor on silicate-rich
alumina substrate (after grinding)

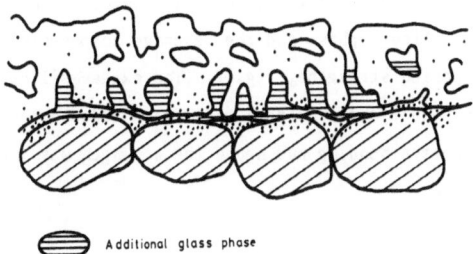

Additional glass phase

Fig.7 Schematic of mixed bonded conductor on
alumina substrate

4   MIXED BONDED CONDUCTORS

The high peak firing temperatures required for reactively bonded conductors led to the introduction of mixed bonded conductor inks which contained glasses or glass formers as well as copper and cadmium. These inks were claimed to combine the best features of fritted and reactively bonded conductors and to be capable of being fired over a wide temperature range, from 850°C to 1000°C.

A conductor typical of these mixed bonded inks is DP 9791 and this was fired at peak temperatures of 850°C and 925°C on the three substrate types and the adhesion strength measured, Table 10. On the Alsimag and Coors substrates, the adhesion strengths of the conductors fired at 850°C were considerably lower than those fired at 925°C. On the Andermann and Ryder substrates there was no obvious effect of firing temperature but the values at 925°C were lower than on the other substrates.

Table 10:
Adhesion strengths of DP 9791 on alumina substrates

Peak Firing Temperature	Alsimag	Coors 995	A and R 974
850°C	2.06	2.55	2.96
925°C	4.52	3.69	2.89

The composition of DP 9791 was as expected, a mixture of reactive components and a lead boro-silicate glass, Table 11. The loss on ignition at 400°C was the organic binder, probably ethyl cellulose. At the lower peak firing temperature the 'reactively bonded' part of the adhesion mechanism does not occur, and the conductor behaves as a fritted conductor might. At the higher temperature the combination of the interaction between the conductor glass and the substrate silicate phase and the reactive components with the alumina provide high adhesion strengths, fig. 7. The excess silicate at the surface of the Andermann and Ryder substrate once again reduced the effectiveness of the reactive bonding and hence the strength was independent of peak firing temperature.

Table 11:
Composition of dried DP 9791

	Weight %
Gold	94.6
Cadmium	0.7
Copper	0.2
Lead Oxide	1.0
Boron Oxide)	
Silicon Dioxide)	1.5
Loss on Ignition	2.0

5   CONCLUSIONS

The adhesion of thick film conductors to alumina substrates is highly
dependent upon the interaction of deliberate additives in the conductor ink
with the materials of the substrate. The bond between these additives and the
substrate is relatively easily made. Control on the level of interaction is
critical to ensure that there is sufficient adhesion between the conductor
metal and the additives.

The mechanical interlock between the conductor metal and glass in
fritted conductors when properly formed is surprisingly strong. Both
incorrect firing conditions and excessive interactions with the substrate can
significantly reduce this mechanical bond. The newer ink compositions
incorporating partially devitrifiable glasses are less process and substrate
sensitive.

The reactively bonded conductors, in principle, offer the best form of
bond between noble metal and substrate - an oxygen metal bond. The high
firing temperatures required for the chemical reaction to occur make
processing more difficult. The propensity of the copper to diffuse out of the
gold and oxidise at the top surface of the conductor at lower temperatures,
such as for overglaze profiles, could present a problem for assembly,
especially wire bonding. Although presented as substrate independent,
reactively bonded conductors are also subject to problems associated with
excessive quantities of silicate phases at the substrate top surfaces.

The mixed bonded conductors offer an attractive solution in combining
the properties of fritted and reactively bonded conductors. Like the
reactively bonded conductors, they seem to be only possible in gold or gold
alloy compositions and also require the high temperature firing for the

reaction to occur. Also, although to a lesser extent, they are susceptible to copper oxide formation and are substrate dependent.

Future developments in conductor ink technology are likely to concentrate on the lower cost silver alloy and nitrogen fireable copper conductors. Although there will no doubt be improvements in glass compositions, it is unlikely that there will be a replacement for the conductor-glass bonding mechanism at least as long as conventional alumina substrates are used with thick film processing.

Acknowledgements

The author thanks Standard Telecommunication Laboratories Limited for permission to publish this paper. Over the years, past and present members of the STL Thick Film Materials Group have contributed to the work and their invaluable help is gratefully acknowledged. Special thanks are due to STC Hybrids Unit, Great Yarmouth for support and expert advice.

References

1. Hoffman, L.C., Bacchetta, V.L. and Frederick, K.W., 'Adhesion of Platinum-Gold Glaze Conductors', Proc. IEEE, Elec. Comp. Conf., 1965, S-381.

2. Crossland, W.A. and Hailes, L., 'Thick Film Conductor Adhesion Reliability', Amer. Cer. Soc., 73rd Annual Meeting, April 1971.

3. Woolley, J.F., 'The Determination of Calcium, Magnesium and Silica in Alumina by Atomic Absorption Spectroscopy', 3rd International Congress of Atomic Fluorescence and Atomic Absorption Spectrometry, September 1981.

4. Coleman, M.V. and Gurnett, G.E., 'Surface Area, Structure and Composition of Debased Alumina Substrates', Proc. IERE/ISHM Conference on Hybrid Microelectronics, September 1975, p 1.

5. Hoffman, L.C., 'Metallising Compositions Containing Bismuthate Glass-Ceramic Conductor Binder', US Patent 3741780, June 1973.

6. Smith, B.R., 'Composition and Method of Bonding Gold to a Ceramic Substrate and a Bonded Gold Article', US Patent 3799890, March 1974.

7. Smith, B.R., 'Gold Composition for Bonding Gold to a Ceramic Substrate Utilising Copper Oxide and Cadmium Oxide', US Patent 3799891, March 1974.

8. Coleman, M.V. and Gurnett, G.E., 'The Limitations of Reactively-Bonded Thick Film Gold Conductors',Proc. ISHM European Hybrid Microelectronics Conf., May 1977.

Chapter 6

Scanning Laser Acoustic Microscopy with some Applications to Adhesion

W.Arnold and H.Reiter

Fraunhofer-Institute for Non-Destructive Testing
Bldg. 37, University, Saarbrücken, Fed. Rep. Germany

I. INTRODUCTION

Acoustic microscopy became a powerful new tool in non-destructive evaluation (nde) of surfaces and subsurface imaging of optically opaque materials, in imaging of cells in biology, and a vast variety of other applications. Two different techniques have emerged for acoustic microscopy. In the Scanning Acoustic Microscope (SAM) a diffraction limited spot of an acoustic beam is obtained by an acoustic lens in direct analogy to optical microscopy. Imaging is performed by measuring the acoustic reflectance or transmission of an object in the focused spot/1/. In contrast, Scanning Laser Acoustic Microscopy (SLAM) utilizes plane waves. Both methods have the ability to penetrate materials provided the sound absorption is sufficiently small at the employed frequency. Scanning Laser Acoustic Microscopy became commercially available some years ago as a tool for non-destructive evaluation of ceramics, tissue characterization of biological materials, and testing of adhesive bonds. We shall first discuss its principle, and then present applications related to adhesion.

II. PRINCIPLE OF OPERATION

The principle of operation of a Scanning Laser Acoustic Microscope (SLAM) is outlined in fig.1. A sample is insonified under a certain angle with respect to the surface of the sample.

In a homogeneous sample the sound causes a ripple of the surface
from which a laser beam is reflected off. The spatial and tempo-
ral periodic displacements of the surface cause the laser beam
to be partially diffracted and frequency shifted due to the Dop-
pler effect. By a knife edge one diffraction order is blocked

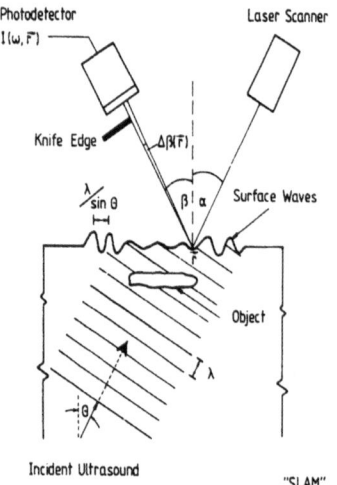

Fig.1: Principle of Scanning
Laser Acoustic Microscopy

which then leads to an ac current in the photodiode because its
output contains a mixing product between the undiffracted zero
order and the still present diffracted part. The frequency of the
ac component is equal to the sound frequency and its magnitude is
proportional to the sound amplitude. It has been shown that such
a technique makes it possible to detect coherent surface waves of
extremly small amplitude, limited only by the shotnoise of the
photodiode/2/. In the presence of pores, inclusions and cracks in
the sample, the sound wave is scattered by these defects which in
turn becomes visible as modulation of the otherwise homogeneous
surface ripple. By scanning the laser beam in a raster manner
over the surface of the sample, this can be measured and displayed on a television screen. A microscope based on this principle is commercially available/3/. It uses sound frequencies of
30, 100 and (optional) 500 MHz. The resolution is given by the
wavelength of the sound wave which is at 100 MHz typically of the
order of 50 μm in most solid materials. The magnification factor
is at 30 MHz ∿25 and at 100 MHz ∿80. Very often one wishes to

examine samples which do not reflect optically, and in this case the sample is covered by a small plastic block. One face is metallized but otherwise the block is optically transparent. This face is laid onto the surface of the sample and the acoustic coupling to the surface displacements of the sample is achieved by water. The laser beam is then reflected off from this coverslip. It is obvious that the images obtained by the SLAM are acoustic shadow graphs of the defects in the sample examined provided their sizes are large compared to the acoustic wavelength λ. If the size becomes comparable to λ, one obtains acoustic diffraction patterns of the defects.

III. APPLICATION OF THE SLAM

A typical application of the SLAM is defect detection in ceramic materials. Ceramic materials made out of Si_3N_4 and SiC became very important as high-strength materials for components in gas-turbines still operable at temperatures as high as 1600 K. Because ceramics are not ductile, they cannot release the stress associated with a defect by plastic deformation, and it must therefore be assured that the components contain no defects larger than a critical size depending on the material and the type of defect/4/. Figure 2 shows an acoustic micrograph of a pore with a diameter of roughly 500 µm imaged at 100 MHz in Si_3N_4.

By a simple electronic circuit, it is possible to use the phase of the rf which excites the transducer as a reference for the phase of the output of the photodiode which in turn is related to the phase of the surface wave displacements. If the samples were homogeneous, the acoustic micrograph would display parallel lines. In areas where the sound attenuation and velocity varies these regular lines are "scrambled", and thus enhance the detectability of a defect (see fig.2)/5/. It should be noted, however, that this is an additional mode of operation of the SLAM (also called interference mode). Furthermore, it is quite simple to determine the depth of a defect. By marking the sample and measuring the apparent distance of the defect to this reference, turning the sample by 180°, measuring this distance again, it is straightforward to deduce the depth by trigonometry (fig.3)/5/.

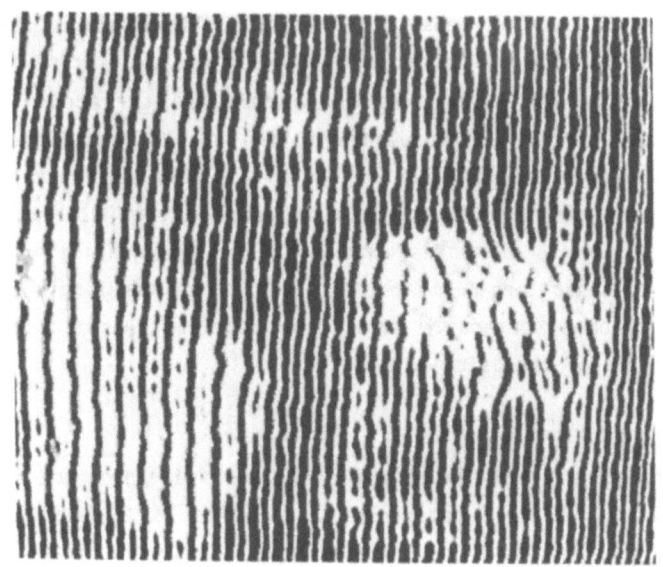

Fig.2: Acoustic micrograph of a pore in the sintered
ceramic material Si_3N_4 . In the interference mode
parallel lines are obtained where the sample is homogeneous whilst a defect results in "scrambled" interference lines. The diameter of the pore is ∿500 μm.

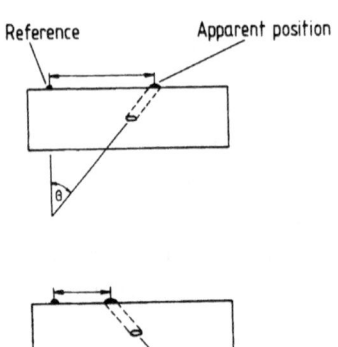

Fig.3 : Principle of depth
determination

In examining adhesive bonds the SLAM can be a powerful tool
to determine their strength. Bonds which are broken will not
transmit ultrasound. Such examples are shown in figures 4 and 5.
Figure 4 displays the honeycomb structure of bonding lines of a

polymer laminate used to increase the visibility of traffic
signs. Good bonds transmit sound effectively (fig. 4b) whereas
poor bonds do not (fig.4a). The micrographs were taken at 30
MHz/6/. Fig. 5 shows an acoustic micrograph of a ceramic chip
capacitor/7/. The condensator consists of several dielectric
layers which are bonded together. One can clearly see that some
of the bonding layers are interrupted. The micrograph was taken
in the interference mode.

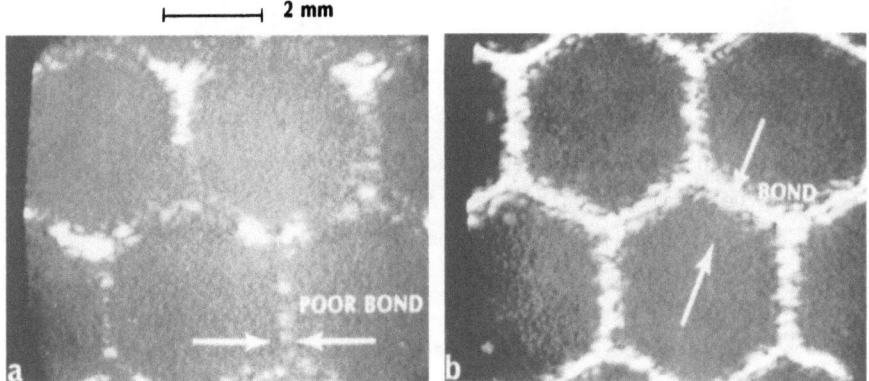

Fig.4: Acoustic micrograph at 30 MHz of the honeycomb
structure of a polymer laminate. Poor bonds do not transmit sound waves (a) whereas good bonds transmit ultrasound
effectively (b) (from ref.6)

Further applications related to bonding have been demonstrated by Sonoscan. We mention specifically seam welding evaluation, epoxy glue testing, and inspection of electronic compo-
nents/6/.

In applications of the SLAM related to adhesion problems we
have mainly examined bonds in solar cells which were obtained by
parallel gap welding. An optical photograph of such a system is
shown in fig.6a. Arrows indicate where the tab is bonded to the
underlying electrode evaporated onto the silicon. There are two
bond lines visible. Acoustic micrograph of the bonding area is
shown in fig.6b. Whereas one bond transmits ultrasound over most

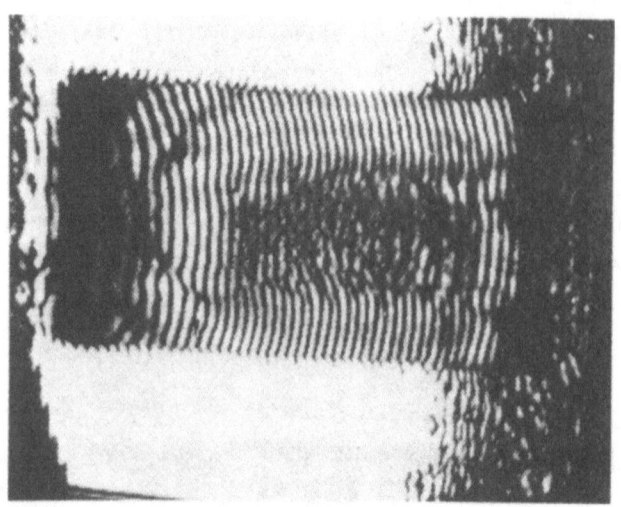

Fig.5: Acoustic micrograph at 100 MHz of a ceramic
capacitor with a layered structure. Delaminations
can be clearly seen by the disturbances of the interference lines (reproduced from ref.6).

of its width, the second one does so only partially. Apparently,
diffusion of the bonding material did not take place over the
whole width of the electrodes (imaged at 100 MHz).

Figure 7 shows an application in diffusion bonding in me-
tallurgy. Two metal sheets made out of titanium were diffusion
bonded over an area of ~ 0.5 m^2. Because of the large ultrasonic
attenuation in this metal at 100 MHz, the specimen was examined
at 30 MHz. It showed areas where the diffusion bonding was defect-
ive as can be seen in fig.7 by arrows.

One of the advantages of the SLAM is the rapid build-up
time of the images. The scanning is carried out in TV-norm thus
rendering possible inspections in real-time. Furthermore, the
acoustic images obtained are from a physical point of view diffraction patterns. It is therefore tempting to restore the images
by corresponding algorithms. Such studies have been undertaken
theoretically/8/ and experimentally along one line of a SLAM

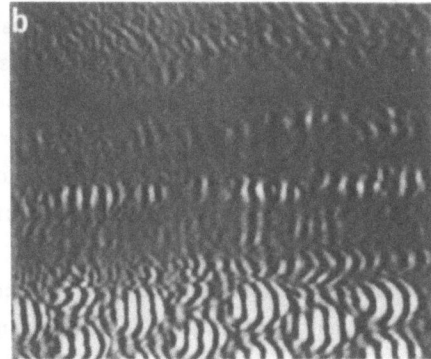

Fig.6a,b: Optical (a) and acoustical (b) image of a
diffusion bond in a solar cell obtained by parallel
gap welding. Whereas one bond is intact over its whole
width, the other one is only partially complete.

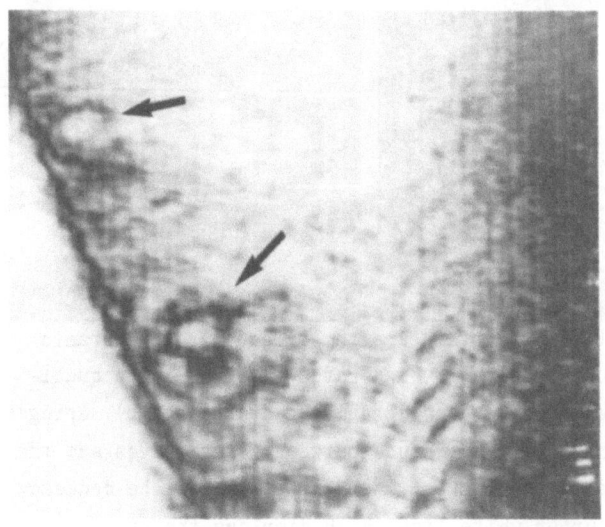

Fig. 7: Acoustic micrographs of diffusion bonded metal
sheets taken at 30 MHz. Defects in the bonds are present
as indicated by the arrows.

picture/9/. A full two-dimensional restoration of the micrographs
has to be carried out on a computer. We are presently working on
this problem. The SLAM was connected to a digital image process-

ing system/10/ with 512 x 512 pixels and 256 grey levels. The image processor itself is linked to a PDP11/34 computer. The SLAM pictures can be stored on floppy disks as well as on a winchester disk. The hardware of the image processing unit contains an averaging procedure which sums 256 frames in about 10 sec. This improves the quality of the SLAM micrographs obtained with poor signal/noise ratios considerably. Such an example is shown in figs. 8a and b. It displays an acoustic micrograph of a pore located at the surface of a hot-pressed SiC part. Whereas in fig.8a noise is degrading the quality of the image, this noise is totally removed in fig. 8b after averaging.

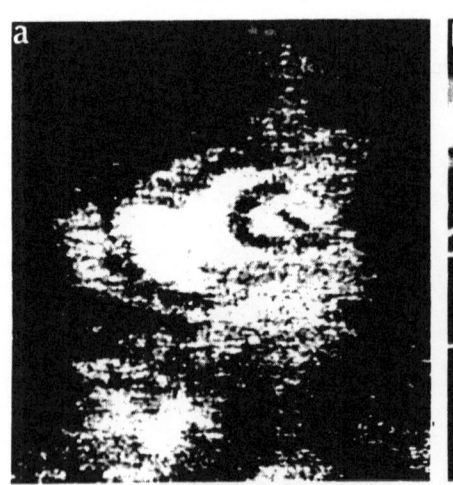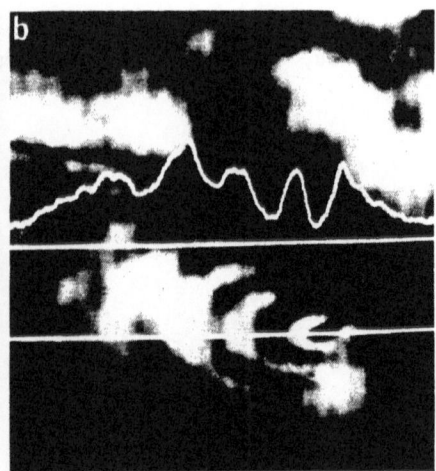

Figs.8a,b: Image obtained of a surface pore in ceramic SiC at 100 MHz with low S/N ratio (a). The image quality can be improved considerably by averaging (b). Fringes appear through interference between volume waves and the scattered surface waves. Their spacing Λ can be deduced from the upper white line which displays the grey level distribution along the lower line (generated by software).

Because the defect is at the surface and thus alters the elastic boundary conditions, some fraction of the incident wave is scattered into surface waves which in turn interfere with the volume waves. This results in fringes whose spacing Λ can be measured in order to deduce the surface-wave velocity provided the angle of incidence θ of the volume waves and their wavelength λ are known/11/. Λ can be expressed as:

$$\Lambda = \left| 1/((\sin\theta/\lambda) - (\cos\psi/\lambda_s)) \right| \qquad (1)$$

Here, ψ is the angle between the direction of the bulk wave dynamic ripple and the scattered surface waves of wavelength λ_s. It is clear that such equations can be solved immediately by corresponding software routines, and thus enhance the versatility of the SLAM. This holds also for image restoration due to the diffraction pattern obtained from volume defects as discussed above.

In summary we should like to point out that the SLAM turned out to be a valuable new tool in nde of small components. The application of the SLAM to test bonds is equally possible. The resolution which can be achieved is of the order of the acoustic wavelength depending on whether the sample is still acoustically transparent at the frequencies employed.

ACKNOWLEDGEMENT

We should like to thank Dr. Kessler from Sonoscan Inc. for permitting to reproduce figs.4 and 5.

REFERENCES:

/1/ Lemons R.A and Quate C.F., Phys. Acoust. (Eds. Thurston R.N. and Mason W.P.) 14, 1, (1979); see also the paper by Nikoonahad M. in this volume

/2/ Whitman R.L. and Korpel A., Appl. Opt. 8, 1567, (1967)

/3/ Sonoscan Inc., Bensenville Il 60106, USA

/4/ Evans, A.G. Proc. NATO-ASI Nitrogen Ceramics, 1981, Brighton (UK)

/5/ Yuhas, D.E. and Kessler L.W., Proc. Conf. on "Scanning Electron Microscopy", p. 385, (1980, Chicago)

/6/ Application notes of Sonoscan Inc.

/7/ Kessler L.W. and Yuhas D.E., Scanning Electron Micr.,
    1, 555 (1978)

/8/ Chou C.H., Khuri-Yakub B.T., and Kino, G.S. in
    "Acoustical Imaging" (Ed. Wang K.Y.) 9, 357, (1979)

/9/ Yuhas D.E., Oravecz M.G., and Kessler L.W., Proc. IEEE
    Ultrasonics Symposium,(Ed. McAvoy B.R.) p.541, (1981)

/10/Imago, produced by Compulog, Böblingen, FRG

/11/Yuhas, D.E. Proc. 1st Int. Symp. Ultrasonics Materials
    Characterization, Nat. Bur. Stand. Publ. No. 596
    (Eds. Berger M. and Linzer M.), p. 357 (1980)

Chapter 7

EPOXY ADHESIVE FORMULATION: ITS INFLUENCE ON CIVIL ENGINEERING
PERFORMANCE

R J LARK* and G C MAYS†

*Jamieson Mackay and Partners, Glasgow, Scotland
†Wolfson Bridge Research Unit, Dundee University, Scotland

1 INTRODUCTION

In civil engineering many of the situations in which adhesives are
used are not truly structural in the sense that the adhesive is not required
to transmit significant shear or tensile stresses. In new construction, for
example, applications might include the use of resins to form bridge
expansion joint nosings, to bed down bridge bearings or as slurry type road
surfaces on either steel or concrete substrates. The application of
'structural' adhesives in new construction is rare, although increasing use
is being made of epoxies as gap fillers between units in segmental precast
prestressed bridge construction.

The use of resins in the repair or strengthening of concrete
structures is more widespread. Epoxies are preferred for bonding new
concrete to old since they have a particularly high tolerance to the
alkaline nature of the fresh concrete. Low viscosity resins are now
commonly employed to inject cracked concrete in order to restore it to
something approaching its original condition. The most interesting development, however, is the use of two part, cold-curing epoxy resin adhesives to
form structural joints between steel and concrete for the strengthening or
repair of existing civil engineering structures. Such surface reinforcement can be used to enhance either the flexural or the shear capacity of a
structure.[1] In either case the local strengthening may be required in the
short term to support the passage of an abnormal load or for more longer
term service under additional design imposed loading.

A parallel technique, termed open sandwich construction, has been

developed at the University of Dundee[2] for new construction. This involves
bonding freshly placed wet concrete to a prepared steel plate. The resulting precast concrete unit is about 30% lighter in weight than a conventional slab reinforced with steel bars. When used to form the roadway deck of
medium span, steel/concrete composite highway bridges, superstructure cost
savings of at least 10% may be achieved. It is this latter development
which has prompted the research which is described in this paper, although
the results may be usefully applied to the many civil engineering applications previously described.

Much of the enthusiasm for resins in civil engineering arises from
their continued and successful development for use in the aerospace and,
more recently, the vehicle industry. Material properties have been
thoroughly researched and there is over twenty years of service experience
with modern aircraft. Unfortunately the experience gained can give no
more than an indication of what can be expected of bonded connections in
civil engineering because different types of adhesive are used, working
environments are different and workmanship under the conditions prevalent
on construction sites is less amenable to quality control than that which
can be obtained in factories. For example, a wide range of environments
can occur in highway bridges with temperatures in steel box girders
varying between extremes of -25^oC and $+70^oC$ and with very high humidites.
Fatigue may arise in bridges due to repetitions of wheel loads and
components in decks can experience up to 7×10^8 cycles due to loading by
axles of commercial vehicles. Of particular importance is that highway
bridges in the UK are assessed for lives of 120 years whereas the
operational lives of aeroframes may be only about 20 years.

The long endurance fatigue performance and some durability aspects
of steel lap joints made with adhesives typical of those used in civil
engineering have already been investigated[3,4]. This paper describes
research undertaken to study experimentally the engineering properties
of these adhesives, the relationship between these properties and their
chemical composition and to determine the time, temperature and moisture
dependence of these properties. From this study a comparison between the
performance of the selected adhesives with that of polymers in general
has been sought.

2 SELECTION OF ADHESIVES FOR STUDY

While recognizing that for any particular application many possible types of adhesive are available, in general the thermosetting adhesives are the most suitable for civil engineering applications. Of these, for structural uses, the epoxies are probably the most versatile and thus to date the two-part cold cure epoxies have invariably been the choice of the construction industry. A general classification of the base resin and hardeners used in such formulations has been carried out[5]. The most commonly occuring resins are the diglycidyl ethers of Bisphenol 'A' and Bisphenol 'F', while hardeners may be polyamides, polyamines of the aliphatic or aromatic variety or polysulphides. Thus, based upon a searching series of preliminary tests, the following adhesives were selected for detailed study. They are referred to by Wolfson Bridge Research Unit designation numbers and the omitted numbers usually represent adhesives rejected from the full testing programme. In all cases the formulations are based upon Bisphenol 'A' or 'F' resins and it is the hardener type which is the variable. It is recognised, however, that the presence of other additions such as fillers, diluents, plasticisers etc., may contribute significantly to the final mechanical properties of the hardened adhesive.

Adhesive No. 1: a white, thixotropic filled paste mixed with 10% of its own weight of a black liquid aliphatic polyamine hardener.

Adhesive No. 2: a high viscosity white, liquid resin mixed with its own weight of an amber coloured, high viscosity liquid polyamide hardener.

Adhesive No. 4: a white, filled paste mixed with approximately half its own weight of a medium viscosity, amber coloured liquid hardener of the aromatic polyamine type.

Adhesive No. 6: a white, filled paste mixed with one third of its own weight of a black aliphatic polyamine adduct hardener paste.

Adhesive No.14: a white, filled paste mixed with an equal volume of a black polysulphide hardener.

3 ADHESIVE PROPERTIES AND THEIR MEASUREMENT

Those engineering properties of the hardened adhesive in an unconfined state which were considered for measurement included tensile, compressive, shear and flexural strengths and fracture toughness. Also included was a test for the heat deflection temperature. The most reliable methods were found to be, in order of preference:

(1)   Four point bending of an adhesive prism for the determination of the flexural modulus of the material (see fig. 1). The specimen size of 200 x 25 x 12 mm was found to be the most convenient for ease of casting, demoulding, versatility for other tests and because it could be tested in a load range suited to the larger test machines found in civil engineering laboratories. The flexural modulus is quoted as the secant modulus at 0.2% strain.

(2)   Heat deflection temperature testing using the same prisms and loading configuration as described in (1) above. The sample under test was placed in a temperature controlled cabinet and a load applied such that the specimen was subjected to a maximum fibre stress of 1.81 N/mm^2 in accordance with BS 2782. The Heat Deflection Temperature (HDT) of the adhesive was taken as the temp-erature, measured on a thermocouple attached to the specimen, attained after it had undergone a further 0.25 mm deflection while subject to a surface heating rate of $0.5^{\circ}C$/minute.

(3)   Bulk shear strength testing of the same prism as in (1) above tested in the shear box illustrated in fig. 2.

Tensile tests and bulk adhesive fracture toughness measurements were found to be unreliable because of the difficulty in producing flaw free specimens. Compressive testing was discontinued because results were both unreliable and of little relevance to this study.

In addition, two tests for adhesion using steel to steel joints were included:

(4)   Joint fracture toughness testing using the tapered double canti-lever specimens shown in fig. 3.

(5)   Double lap shear joint testing with the specimen configuration given in fig.4.

A standard steel surface preparation procedure involving degreasing with detergent solution and grit blasting to Swedish Grade Sa2½[6] was employed throughout. The measurement of lap shear strength, although somewhat variable and sensitive to surface preparation standards, was retained because it is a standard test and therefore useful for comparative purposes.

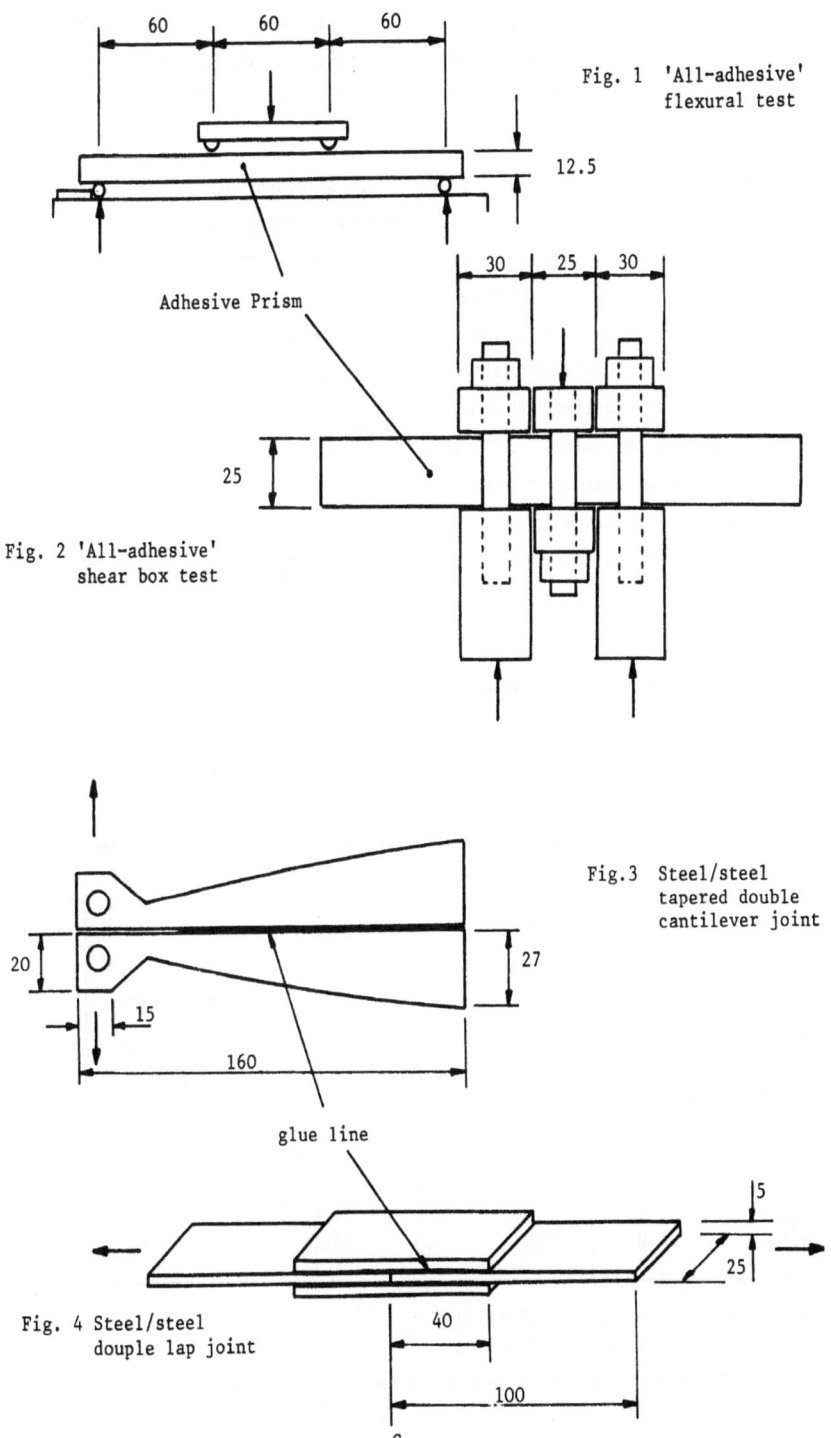

Fig. 1 'All-adhesive' flexural test

Adhesive Prism

Fig. 2 'All-adhesive' shear box test

Fig.3 Steel/steel tapered double cantilever joint

glue line

Fig. 4 Steel/steel douple lap joint

Steel-to-concrete joints were not used as failure of such specimens generally occurs within the concrete and little information is provided on the
performance of the adhesive itself or of the adherend/adhesive interface.

3.1   Results and Discussion

The results of the mechanical tests on the selected adhesives are
summarised in table 1. It should be noted that with adhesive No.14,
specimens were tested after 28 days curing at 20°C rather than the custom-
ary 7 days. This was found to be necessary because of the extremely slow
cure rate of this adhesive. At first, it was necessary to have the grit
blasting of steel for joints carried out by an outside contractor causing
a delay between grit blasting and manufacture of the joints. These are
referred to as type 'A' joints in the table. Later, in-house grit blasting
became possible which enabled specimens to be cleaned, blasted and cast in
one day. These are referred to as type 'B'. The relative sensitivity of
the various formulations to surface preparation standards can be clearly
seen.

Table 1:

Mechanical properties of adhesives

Property \ Adhesive No.	1	2	4	6	14
Bulk hardened adhesive					
Flexural modulus (N/mm^2)	8400	2100	3100	7800	2100
Shear strength (N/mm^2)	36.5	15.0	27.9	24.3	18.5
H.D.T. $^{\circ}$C	41	40	48	43	34
Steel/steel joints					
Fracture toughness) A ($MN\ m^{-3/2}$ or $MPa\sqrt{m}$)	1.1	0.5	0.5	2.0	-
) B	2.7	0.5	0.8	1.9	1.0
Lap Shear) A (N/mm^2)	10.3	9.0	11.5	18.8	-
) B	24.6	17.0	13.4	21.2	14.1

NOTES: 1.   A = external grit blasting
           B = in-house grit blasting

       2.   All results based on a mean of at least 2 specimens

Adhesive No. 1 has the highest bulk shear strength and flexural modulus of all the adhesives tested and, as might be expected of an epoxy polyamine, it is a fast reacting formulation. Some cohesive joint failures were obtained when used with freshly grit blasted steel but after external blasting failures usually occurred at the interface. In contrast, adhesive No. 2, the epoxy polyamide, has a long pot life but is a low strength, low modulus adhesive, exhibiting only a limited degree of toughness. Inspection of the joint surfaces after failure revealed that the adhesive had undergone considerable distortion and was severely cracked. Adhesive No. 4, although relatively flexible as defined by its flexural modulus, has a degree of toughness only marginally greater than that of adhesive No. 2. Bulk strengths are, however, high as is typical of adhesives of this type. Joint failure modes were generally adhesive in character. The mechanical properties of adhesive No. 6 are similar to those of adhesive No. 1, confirming the belief that these adhesives are of a similar composition. With adhesive No. 6, however, all lap joint failure modes were at least partially cohesive and tapered double cantilever failure modes were entirely cohesive. Lap shear strengths and joint fracture toughness values were little affected by the different surface preparation procedures used. Adhesive No. 14, with a polysulphide hardener, is a low strength, low modulus adhesive exhibiting a medium degree of toughness. Joint failure modes were generally cohesive and deformations at failure were observed to be large. The H.D.T. of this adhesive is at least 20% lower than that of the other adhesives tested and it is below the maximum temperature of 38°C possible on the soffit of a concrete bridge deck.

All these results are in line with the general observations of the review of the chemical composition of epoxies referred to earlier[5], and they confirm the view that the properties of these adhesives can be related to variations in their basic formulation.

4   TIME, TEMPERATURE AND MOISTURE DEPENDENCE

4.1   Time dependency

The time dependency of adhesives Nos 2,4,6 and 14 was monitored by subjecting prisms of each adhesive to a series of sustained loads. The specimen size and test configuration used was the same as that described earlier for the static flexural test. Four beams of each adhesive were

loaded to four different levels resulting in extreme fibre stresses varying
from 0.25 N/mm^2 to 2.0 N/mm^2. Beam deflections, monitored over a period of
at least one year and still continuing, have then been used to produce
curves representing the effective decay in relaxation modulus of the
adhesives with time. These are reproduced in fig. 5. They represent the
stability of the adhesives with time. In this respect a flatter curve is
beneficial, although the long term low values of the effective modulus of
adhesives Nos 2 and 14 does give some cause for concern as to their
potential structural efficiency under sustained load.

4.2   Temperature dependency

The temperature dependency of the mechanical properties of all five
adhesives was investigated by testing within a temperature controlled
cabinet. The properties studied were the bulk adhesive flexural modulus,
bulk adhesive shear strength, steel double lap shear strength and steel/
adhesive joint Mode I strain energy release rate. The latter is related to
the fracture toughness of the adhesive by its modulus, but because the
modulus is itself temperature dependent, results are expressed in terms of
the strain energy release rate.

The variation of these properties with adhesive surface temperature
at the time of test, measured using a small thermocouple attached to the
exposed adhesive face, is shown in figs. 6 to 9, respectively. From these
results it is evident that the response of all five adhesives to
temperature variations within the range 15° to 65°C is very similar. The
most noticeable feature of the curves is the rapid deterioration which all
properties exhibit at a temperature close to that defined as the H.D.T. of
the material. Around this temperature the failure modes of the jointed
specimens also changed from partly cohesive to adhesive with large
deformations.

This general correlation between the H.D.T. of the adhesives and the
temperature range within which their basic engineering properties undergo
significant change, confirms the proposition that the H.D.T. of an adhesive
is associated with a temperature at which important changes in the
molecular structure of the adhesive are occurring and which affect the way
in which they are capable of carrying load. This is what would be expected
of polymers in general and it supports the view that with respect to
temperature, even these highly modified epoxies behave in a manner typical
of all polymers.

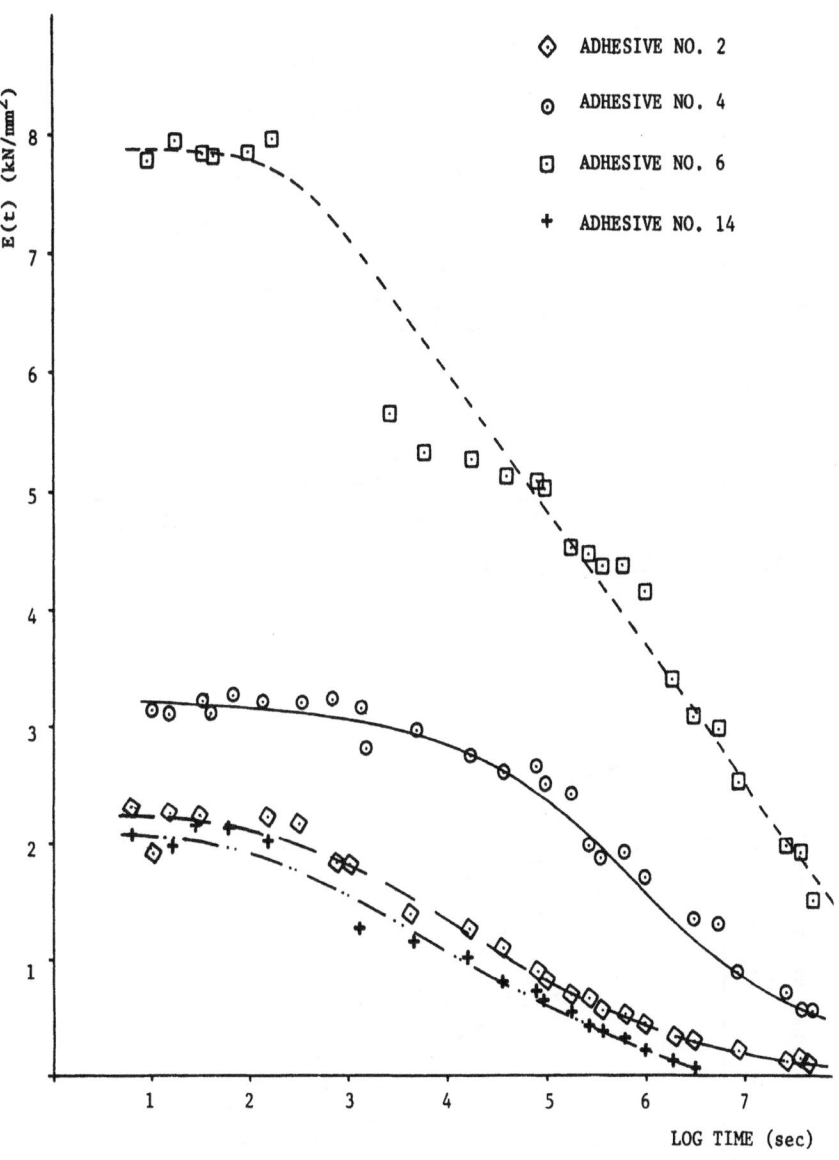

Fig. 5   The variation of the relaxation moduli of
         adhesives Nos 2, 4, 6 and 14 with time

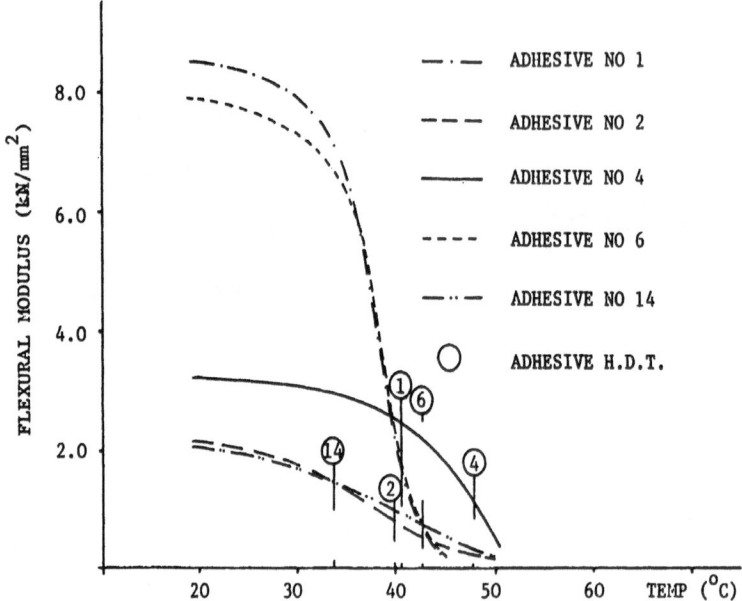

Fig. 6   Temperature dependence of bulk adhesive flexural modulus

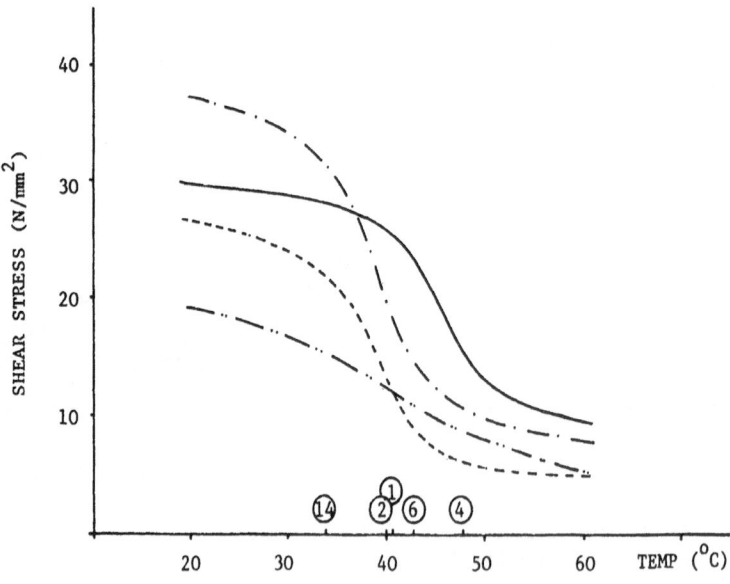

Fig. 7   Temperature dependence of bulk adhesive shear strength

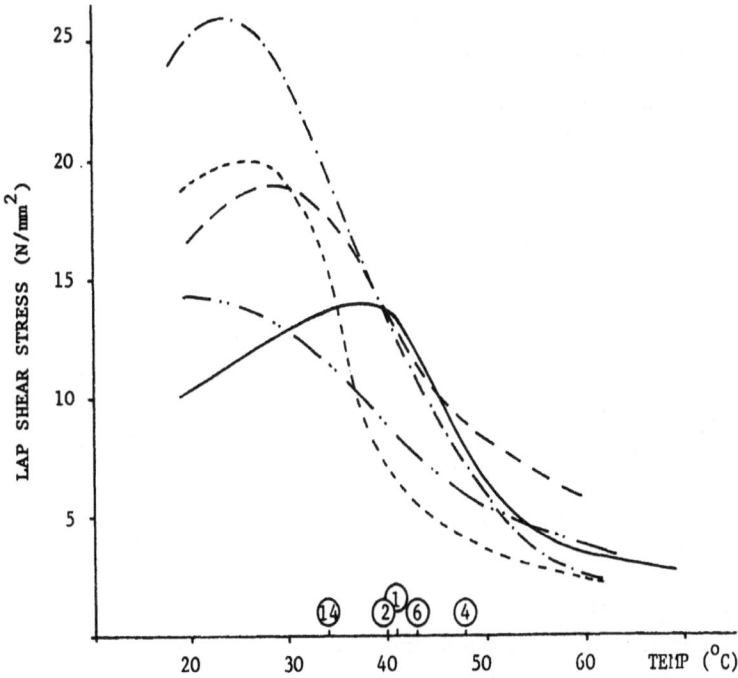

Fig. 8   Temperature dependence of steel/adhesive lap shear strengths

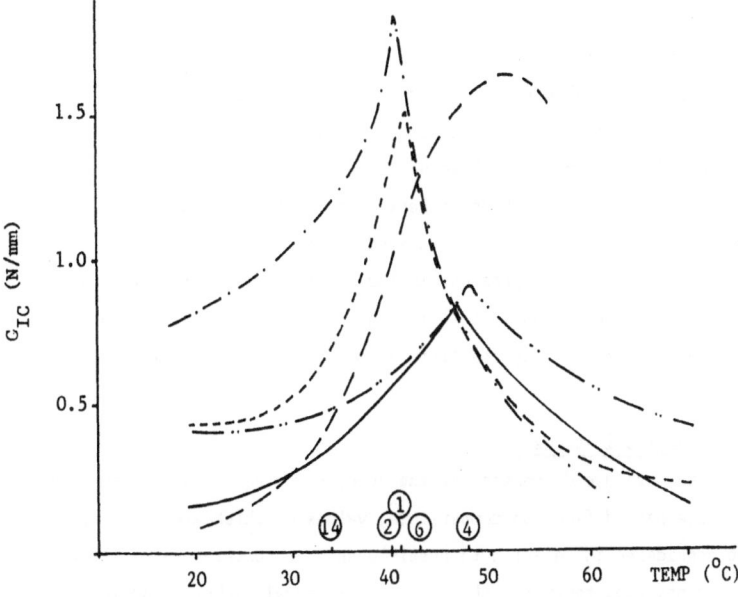

Fig. 9   Temperature dependence of steel/adhesive joint mode 1
strain energy release rate

The broad maxima in the lap shear strength against temperature curves are thought to be the result of three effects:

(i)   the general strength-temperature relationship as expressed by the bulk shear strength of the adhesive,

(ii)   differential expansion of the adhesive and the steel, and

(iii)   relaxation of the adhesive at above ambient temperatures modifying the tensile stress at the ends of the lap joint.

The reasons for the peaks in the strain energy release rate/ temperature curves are believed to be that up to the H.D.T., the increasing flexibility of the adhesives allows them to deform at the crack tip, redistributing the load, and hence increasing their toughness. Beyond the H.D.T., however, although the flexibility of the material is still increasing, it does not fully compensate for the rapid decrease in the material and bond strengths which occur.

4.3   Moisture dependency

To examine the moisture dependence of the adhesives used in this study prisms of size 60 x 12 x 2 mm were completely immersed in a water bath at room temperature (15-18°C). The smaller prisms were adopted to enable results for water uptake to be obtained within a reasonable period of time. Results are presented in fig. 10 and from them it is evident that the relatively unmodified polyamide (adhesive No 2) and the aliphatic poly-amine (adhesive No 1) absorb a significant quantity of water (greater than 5%) while the aromatic polyamine (adhesive No 4) and the adducted, and therefore modified, aliphatic polyamine (adhesive No 6) absorb the least (less than 1%). The polysulphide (adhesive No 14) lies in the middle of the range. All five adhesives are weakened by water absorption, as measured by bulk shear strengths on similar specimens (see fig. 11). With the exception of adhesive No 4, this is associated with plasticisation of the adhesives, as measured by the flexural modulus of the specimens (see fig. 12).

4.4   General implications

The temperature dependence of the adhesives studied may be expressed in terms of a heat deflection temperature, values of which are relatively low when considered in relation to possible service temperatures. All the adhesives monitored, however, exhibited some residual strength and stiff-ness at temperatures above this transition, and further work[7] has shown

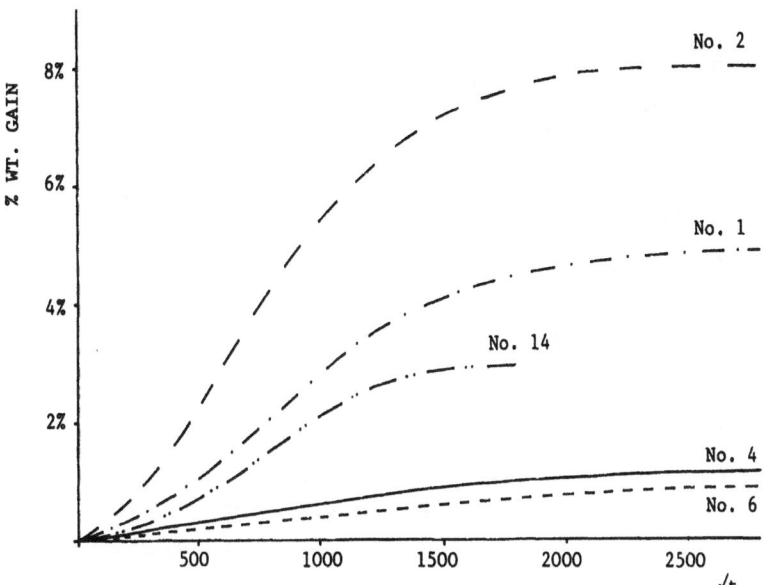

Fig. 10 Water uptake plots for adhesives Nos 1, 2, 4, 6
and 14 - specimen size 60 x 12 x 2 mm

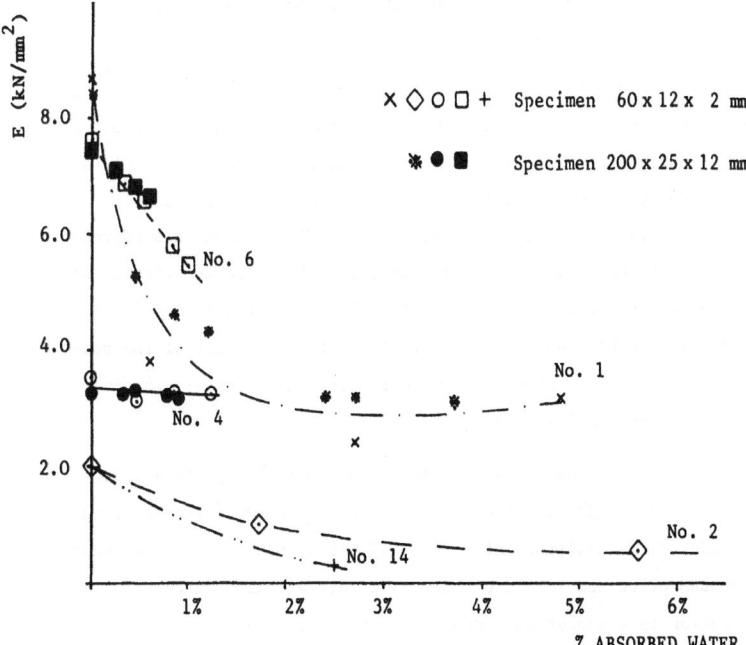

Fig. 11 Experimental relationship between the water absorption of
adhesives Nos 1,2,4,6 and 14 and their flexural moduli

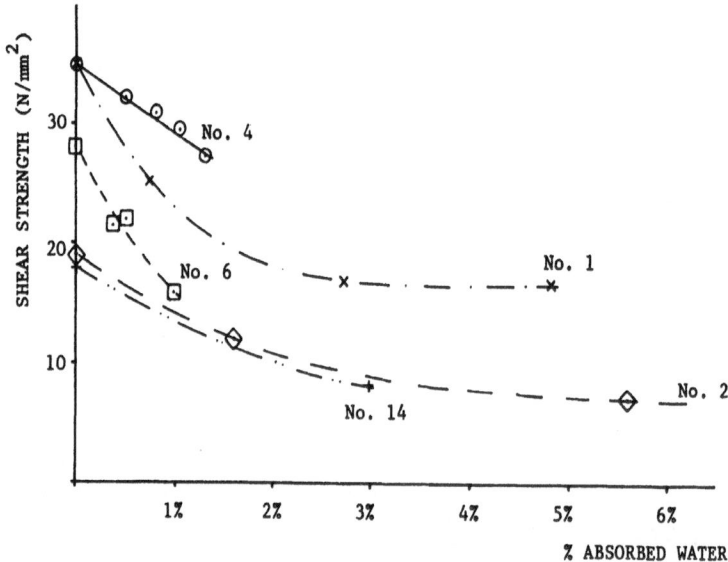

Fig. 12 Experimental relationship between the water absorption of
adhesives Nos 1,2,4,6 and 14 and their bulk shear strength

that open sandwich beams can still perform satisfactorily at these
temperatures.

Comparable observations result from the study of the time
dependence of these adhesives, and in view of the similarity between the
decay of the adhesive moduli with both time and temperature, it would seem
likely that the time/temperature superposition techniques applicable to
polymers in general could also be applied here. Based upon this approach
it is possible to both (i) relate potential long term characteristics of
an adhesive to its short term properties, and (ii) relate the engineering
properties of an adhesive to the more fundamental properties of the polymeric group to which it belongs. A more rational approach to adhesive
selection is therefore feasible.

The sensitivity of the adhesives to moisture absorption suggests
that further links might exist between the short term performance of the
adhesive and its durability. All of the adhesives studied were weakened
by water absorption and in all but one case this was associated with
plasticisation in a manner analogous to their time dependent relaxation.

5 CONCLUSIONS

In line with the general impressions gained during the preparation of a chemical classification of epoxy adhesive formulations, simple mechanical tests have shown that epoxies with aliphatic amine hardeners result in high performance adhesives but which, unless suitably modified, are likely to be intolerant to adverse conditions. The epoxy polysulphide and polyamide are flexible formulations both of which are liable to creep. As the latter also appears to be sensitive to moisture absorption it is unlikely to form the basis of adhesives suitable for structural applications. An adhesive with an aromatic amine hardener gives a durable but brittle adhesive which would benefit from some toughening for structural applications.

The performance of the adhesives used in this study appears to be typical of polymers in general. This is apparent from data obtained using simple, efficient test techniques relevant to the civil engineering industry. From this data it is possible to extract information relevant to both the short and long term performance of the adhesives. The importance of the correlation between these results and what might be expected of polymers in general, is that it allows the Engineer to discriminate between the wide range of materials available in a rational manner. In view of this, the need for the adhesive manufacturer to provide appropriate material properties is evident. Greater discourse between the adhesive suppliers and users within the civil engineering industry is therefore urged.

ACKNOWLEDGEMENTS

The research described in this paper was performed within the Department of Civil Engineering, University of Dundee with financial support from the Science and Engineering Research Council. Adhesives were supplied by Ciba-Geigy Ltd., Permabond Ltd., and Sika Ltd.

REFERENCES

1. Raithby, K.D. External strengthening of concrete bridges with bonded steel plates. TRRL Supplementary Report 612, Crowthorne, 1980.

2. Mays, G.C. and Vardy, A.E. Adhesive bonded steel/concrete composite construction. Int. J. Adhesion & Adhesives, V2, No. 2, April 1982.

3. Mays, G.C. and Tilly, G.P. Long endurance fatigue performance of bonded structural joints. Int. J. Adhesion & Adhesives, V2, No. 2, April 1982.

4. Mays, G.C. and Vardy, A.E. Fatigue performance and durability of epoxy resin bonded metal lap joints. Fatigue in Polymers, PRI, London, 1983.

5. Lark, R.J. Epoxy resin adhesives, a chemical classification. Wolfson Bridge Research Unit, Interim Report WBRU/IR 34 Rev A, Dundee, 1983.

6. Swedish Standard SIS 05 59 00 - 1967. Pictorial surface preparation standards for painting steel surfaces. Swedish Standards Institution, Stockholm, 1967.

7. Lark, R.J. The classification and control of epoxy adhesives in civil engineering PhD thesis Dundee University, November 1983.

Chapter 8

STRUCTURALLY BONDED VEHICLES:
DESIGN AND PRODUCTION CONSIDERATIONS
IMPLICATIONS FOR ADHESIVE FORMULATION

W.A. Lees

Technical Director, Permabond Adhesives Limited, Eastleigh,
Hampshire, SO5 4EX

INTRODUCTION

There is a considerable dilemma within the road vehicle industry concerning
the identify of the materials from which the structures of future vehicles
will be manufactured. Current favourites are aluminium and high strength
steel and it is proposed that bodies fabricated from these metals will be
clad and trimmed to varying degrees with a variety of thermoplastics -
often of the RIM and RRIM types - with the thermoset composites playing,
for the time being, a secondary role.

However, one thing is quite clear - adhesives are going to play a fundamental role in the construction of such vehicles. The reasons are not hard to
find. Aluminium can be very difficult to weld and thin gauge 'high strength'
steels bring their own problems - not the least of which is corrosion.

In both cases these difficulties - and others - can be overcome by utilising
high performance adhesives of the 'Toughened' type. Similarly, the assembly
of composites benefits enormously from the use of such adhesives since bonded
joints avoid the point loads inherently induced by mechanical fasteners.

The extent to which structural adhesives are likely to be used in the construction of future designs will depend not only upon the performance of the adhesives themselves but also upon the appreciation by the designer and production engineer of their capabilities and the technical and economic benefits they can offer. These may be readily summarised and are presented in Fig. 1.

The difficulties are not so readily illustrated but the general problem faced by all concerned is the balancing of the designer's requirements with the adhesive manufacturer's capabilities, a problem compounded by the general influence of an ever shifting and expanding material's technology.

The purpose of this paper is to examine the various issues which are involved in the delicate balance between what is needed and what is possible.

HISTORY

High performance adhesives have been used in the aircraft industry for some forty years, but these materials - based on modified phenolic adhesives (1) - are quite unsuitable for structural applications in road vehicles. However, the introduction by Goodrich (2) and Dupont (3) of various means of 'Toughening' epoxy and acrylic based adhesives led to the early development by Permabond (4) of a number of these novel adhesives which were intrinsically capable of meeting the needs of mass production.

1974 saw the bonding of hinges, locks and door panels of the SMC based ERF lorry. Using the first two-part, cold cured toughened epoxide (5). The first commercial single-part epoxide (5) was introduced in 1975 - a developed version of this adhesive was used in the construction of the bonded aluminium monocoque body of the ECV-3 by British Leyland Technology (6).

These and many other applications for toughened adhesives - not forgetting the acrylic variants (7) - have led to a burgeoning interest in the structural bonding of vehicles. An interest which is expressed in the work being carried out at Harwell (UK - AERE) under the auspices of the composite/metal jointing working party - supported by the EEC. Here, attention is also being paid to the capability of adhesives to deliver power - through bonded joints. The limited information published so far (8) indicates that this should be possible and that the pioneering work in this area by Ford (9) has been substantially advanced.

However, the foregoing should be seen in the correct context - an indication of future design trends rather than a measure of immediate commercial significance. For the present, we are likely to see an increasing consumption of structural adhesives in their 'weld-bonding' role - where adhesives are used to supplement the performance of welded joints. It is noteworthy that this technique,

113

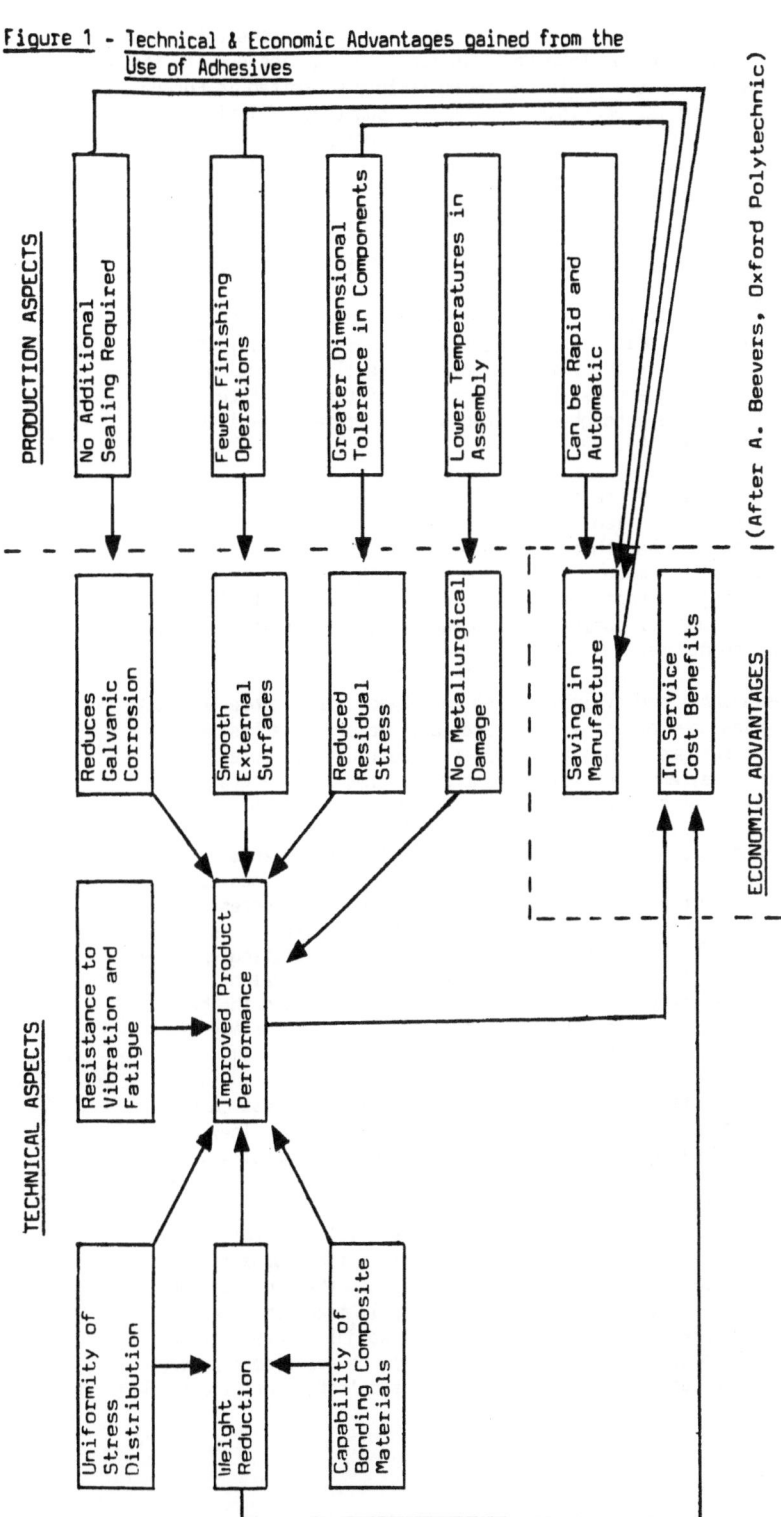

Figure 1 - Technical & Economic Advantages gained from the Use of Adhesives

(After A. Beevers, Oxford Polytechnic)

originally developed by the Russians (10), has been extensively taken up by the Japanese Motor Industry - initially using high quality though conventional epoxides. It is now being rapidly exploited in Europe where both the normal epoxides and the new toughened variants are involved. This is seen to be an excellent development for it increases the vehicle manufacturers awareness of adhesive technology - its capabilities as well as its shortcomings - without requiring the commitment that total reliance on adhesives would need.

THE NATURE OF MODERN STRUCTURAL ADHESIVES - DESIGN IMPLICATIONS

There is a substantial literature devoted to adhesion science though a simple overview, which also considers why adhesives fail, is available (7). While the latter discusses the 'Toughening' process briefly, a thorough and extensive account - of the use of rubbery polymers to produce a toughened two-phase structure - is given by Kinloch and Young (11).

A detailed summary of the characteristics of the various toughened adhesives is given in Table 1. Though for the present, it is sufficient to say that these modern materials have one outstanding characteristic in common - they are robust. They do not crack readily and are, with few exceptions, particularly resistant to their environment - whether during the manufacturing process itself or in their subsequent use.

Crack resistance; and by implication an enhanced resistance to shock, peel and cleavage forces; is a natural outcome of the adhesive's toughened structure. This also supports their overall durability since high Tg's, and therefore enhanced performance at elevated temperatures, can be maintained without inducing brittleness. Furthermore, their excellent strain to failure ratios ensure that moisture absorption is not too disruptive. But, the acknowledged capacity of toughened adhesives to cope with poorly cleaned surfaces is more a measure of the formulator's skill - than an inherent property of the materials themselves. In this context, it is perhaps worth pointing out that the toughened two-part cold setting epoxides are still subject to the same restraints as their equivalent untoughened versions. Unlike the toughened acrylics and the heat cured toughened epoxides, the two-part cold cured systems may be badly affected by oil contaminated surfaces.

It is for this reason that the toughened acrylics and heat cured epoxides have gained the greatest degree of popularity. They are generally acknowledged to be 'strong', durable and generally very easy to use. Despite this, it should be realised by those who have not met them before that these adhesives - unlike the traditional types whose history goes back at least forty years - are still in a relatively early stage of their development. Despite this the toughened adhesives can be truly claimed to be better than their untoughened predecessors. For the designer, this generally means that early concepts of double, strapped, capped or tapered joints are no longer required. Such assembly techniques need only be reserved for power delivery and occasional use in highly stressed 'safety-critical' areas.

Table 1 - General Characteristics of Current Toughened Adhesives

ACRYLIC

Two-part mixed: Usually mixed in ratio 1 : 1 or 1 : 2. General purpose gap filling adhesives. Overall performance not usually as good as non-mixed versions below.

Two-part - not mixed: One component is painted onto surface to be bonded which it catalyses. Adhesive part then applied. Limited to relatively thin bond lines.

Viscosity: viscous liquids to thixotropic pastes.

Strength:[+]   shear - up to 35 MPa
              T-peel - usually 100-120 N/25 mm width
              impact - stops standard ASTM pendulum.

Cure Time: from 15 seconds to 5 minutes according to
$(25°C)$ formulation.

EPOXY

Single part: Heat cured high performance adhesives with general utility in both mechanical and structural applications. Particularly suited to the harsher environments.

Viscosity: Thixotropic gels to stiff pastes - note, some are
             intended to flow like solder when hot.

Strength:[+]   As above (Acrylic)

Cure Time: From 1 minute according to material, method and
             temperature - usually about 30 minutes at $180°C$.

Two part: General purpose high performance adhesives with good environmental resistance. Variety of mix ratios.

Viscosity: Low viscosity resins to thick pastes.

Strength:[+]   As above (Acrylic)

Cure Time: Not less than 2 hours, unless warmed.
$(25°C)$

[+]   See also Table 2.

However, in these latter situations, the concept of 'strength' as measured by standard lap joint and related techniques is simply not good enough. Standard test methods cannot give any indication of how the adhesive in question will respond to either a change in the adherends or a change in the geometry of the joint. Furthermore, the performance observed cannot be extrapolated to that of a full size structure.

An insight into the performance of real bonded structures may only be obtained from an appreciation of the 'Engineering Characteristics' of the adhesive concerned and their interaction with a mathematical model of the structure itself. The appropriate characteristics - Shear Modulus, Elastic Shear Strain Limit and Asymptotic Shear Stress are probably best obtained from the Thick Adherend Shear Test (TAST) (12). This knowledge, coupled with modelling techniques such as Finite Element Analysis can provide the designer with a very powerful tool and a profound insight into what is likely to happen (13,14).

It is clear from such work that in order to cope with a wide range of applications an equally wide range of characteristics is required. Simple concepts of strength are not enough. Consider the following example which relates to two toughened adhesives - both of which are capable of taking standard 1 mm mild steel test pieces beyond their elastic limit. One adhesive, being a toughened acrylic is compliant and flexible while the other, a toughened epoxide, is stiff though not brittle. Their contrasting characteristics are summarised in Table 2 while Figs. 2 and 3 compare the stress patterns which simple but effective modelling predicts for a given load on steel/steel and steel/composite joints. It is very interesting to note that these adhesives, when judged conventionally, have similar 'strength' characteristics, but differ fundamentally in their practical performance when assessed properly. In Fig. 2 - the composite/steel joint - the asymmetric load distribution is less marked in the case of the acrylic based joint because the more compliant adhesive distributes the load more readily and over a greater area than the stiffer epoxide.

Table 2 - Further Properties of Typical Toughened Adhesives (25°C)

	Acrylic Base	Single Part Epoxide
Shear Modulus (GPa)	0.35	1.82
Elastic Shear Stress Limit (MPa)	2.25	25.21
Asymptotic Shear Stress (MPa)	24.22	42.54

A similar pattern of load distribution is seen in Fig. 3 - the steel/steel joint. Although the load distribution within the joint is essentially symmetrical - because both adherends have the same modulus - the acrylic adhesive once again distributes the imposed load over a much greater area than the stiffer epoxide.

Thus, other issues apart, it is reasonable to consider that for the composite joint a compliant acrylic adhesive would be better than a stiff epoxide since it would stress the brittle surface of the composite to a lesser extent than the stiffer epoxy adhesive.

Obviously, as the individual components of the bonded structure become more complex, the adhesive/adherend interaction becomes even more compounded. Further consideration of the wide range of candidate materials from which a future vehicle might be built, gives some indication of the problems faced by the adhesive manufacturer. Complex though these are, they can be reduced to two major issues. The low surface energy of some plastics (the wetting problem), which will be dealt with later, and the contrasting stiffness of the various materials where the shear modulus varies by a factor of 1,000 as one progresses from compliant plastics to High Strength steel.

Figure 2

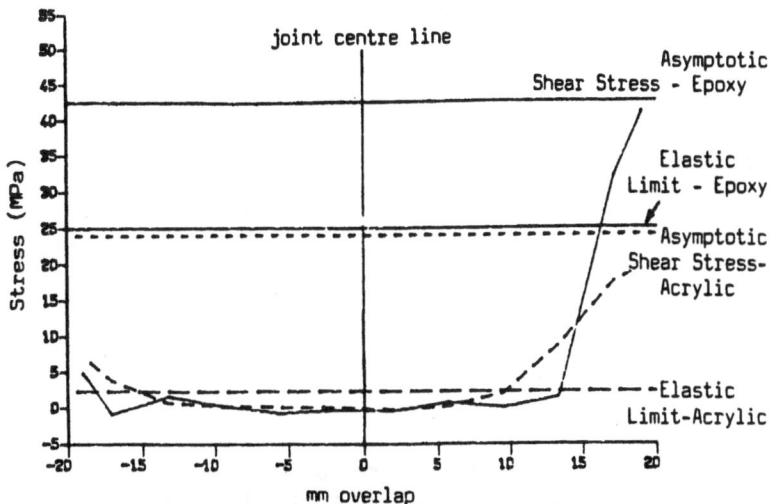

Comparison of stress distribution - for a 10KN load - in two 75mm
wide x 38mm overlap steel (1mm)/composite (1mm) lap joints. Dashed
lines relate to acrylic adhesive ; solid lines to epoxy adhesive
(see Table 2 for properties).

Note: 1) Although large this joint is not untypical of a metal/
          composite joint and allows comparison with later Figures -
          particularly Figure 6.

      2) This static load can be carried but repeated imposition
          would probably case fatigue failure for both adhesives are
          stressed beyond their elastic limit.

Historically, the modulus problem has been dealt with on a purely
empirical basis and it is only recently that advances in modelling
techniques have given an insight into what is really required. In
a word - this is the 'articulation' of the elastic/plastic behaviour
of adhesives shown schematically in Fig. 4 where the two extreme
curves bracket the possibilities generated by the 'articulation' of
the intermediate example.

Formulation based on such a concept - which could well be called
'Adhesive Engineering' - is now well in hand though inevitably some
time must elapse before the vehicle designers will benefit from the
availability of an adhesive range possessing a carefully planned
characterisation matrix.

Figure 3

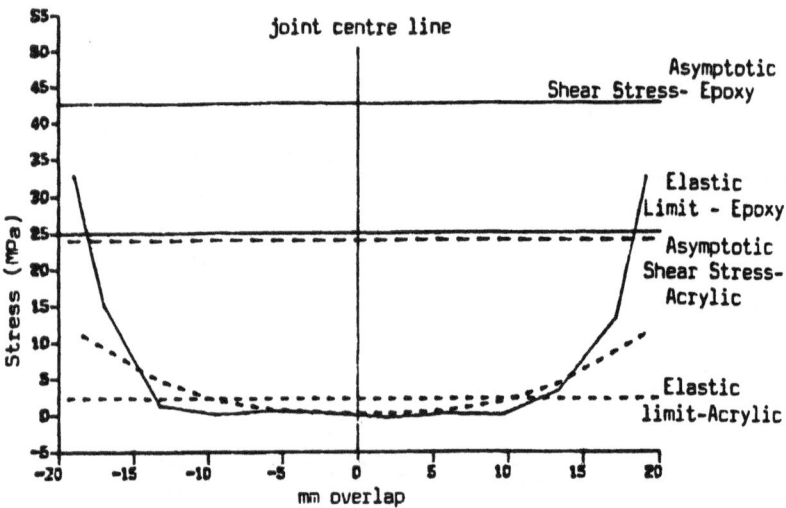

Comparison of stress distribution - for a 10 KN load - in two 75mm
wide x 38mm overlap steel (1mm)/steel. (1mm) lap joints. Dashed lines
relate to acrylic adhesive; solid lines to epoxy adhesive
(see Table 2).

Figure 4

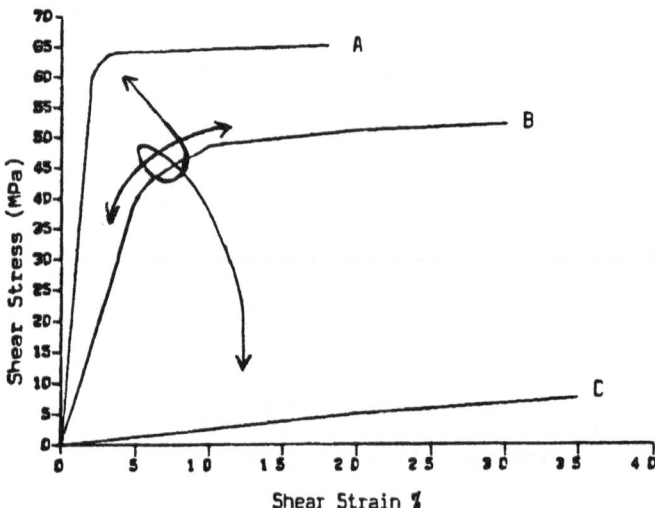

The 'articulation' of the elastic/plastic properties of an adhesive to
give a required performance.

A — maximise stiffness to give good high temperature performance.
B — Typical general purpose material.
C — A characteristic form displayed by many acrylic based adhesives.

In parallel with such developments, the formulator must also ensure that any particular adhesive has the necessary viscosity/thixotropy characteristics coupled with the appropriate rheology properties in order that it may be utilised effectively. During any of the production processes it is likely to experience. A generalised list of all the factors requiring consideration is given in Table 3.

Table 3 - Structural Adhesives - Property Profile for Design and Quality Assurance (see also Table 8)

Primary Properties

Modulus   )
Elastic Shear Stress Limit)   at various temperatures
Asymptotic Shear Stress   )
Glass Transition temperature

Standardised Strength Characteristics

Shear   Impact
Peel   Combined impact/peel

Environmental Performance

Boeing Wedge   Salt Spray
Fatigue   Resistance to shear stress at elevated
Creep   temperatures and humidity

Application Profile

Viscosity   Cure characteristics
Thixotropy   Surface preparation required
Rheology   Physiological activity
Storage requirements

DESIGN ASPECTS AND SPECIAL CONSIDERATIONS

The many interacting benefits offered by the use of adhesives have
already been illustrated in Fig.1 and are summarised in Table 4.
While it is obvious that they comprise a powerful incentive to the
designer to use adhesives, it is readily appreciated that for him to
do so required considerable courage. Not only because of the inherent
risks embedded in a new philosophy but also because the reduction of
welding to a supplementary role will demand new techniques and design
concepts - particularly in such areas as production and repair where
there are enormous investments in 'Know-How', and equipment. It is
fortunate therefore that the technique of 'weld-bonding' is available
for this, coupled with the well established bonding of secondary
structures, can be used as a foundation upon which a greater
commitment to structural bonding may be safely established.

Table 4 - Benefits of Structural Bonding with Toughened Adhesives

Initial Design

* Enhanced stiffness allows use of lighter alloys and/or thinner
  gauges.

* Allows combination of many different materials.

* Avoids high local loads.

* Greater design freedom due to absence of dimensional and access
  restraints induced by conventional welding techniques.

Construction

* No metalurgical damage.

* No distortion or defacement.

* Very accurate assembly possible without undue expense.

* Little or no residual stress.

* Can simplify assembly techniques.

* Capital costs usually lower.

* Overall costs can be reduced.

* Reduced health hazard.

Ultimate Use

* Repair can be simpler than welding.

* No metallurgical damage during repair.

* Corrosion reduced or eliminated.

* Old composites may be repaired well.

* Fatigue resistance enhanced.

To a certain degree the introduction will also be both eased and
required by the predicted (15,16) introduction of much larger metal
pressings and composite components.

From concept to construction the design of a new bonded vehicle and its component parts must pass through the following stages:-

* Definition of general requirements and outline specification.

* Design of body parts and sub-components.

* Specific design of individual joints.

* Assessment of the capability of the various adhesives available to meet the requirements of the designed joints.

* Re-assessment of joint design in the light of the actual performance capability of the relevant adhesives.

* Construction and evaluation of bonded parts, components and body shell.

During each of these several stages the designer must bear in mind the interaction of the relevant aspects of Table 5 where a variety of subjects which require special consideration are identified. The more pertinent of these are reviewed below -

Table 5 - Special Design Considerations

Component and vehicle stiffness

Primary load bearing areas -

      * Energy absorption (both intended use and abuse)

      * Power delivery

Repair.

Adherend problems.

Adhesive problems.

Codes of Practice.

Stiffness

Bonded structures are inherently stiffer than their spot-welded equivalents and in the case of box section beams, the flexural stiffness and ultimate strength can be doubled by changing from a welded to a bonded form of construction (17). Bonded doublers are also very effective. A body shell will reflect the increased stiffness of its sub-components and therefore attention must be paid to the correct relative performance of the 'crumple-zones'.

It has been shown (18) that bonded symmetrical tubular beams of both steel and aluminium alloy will absorb energy just as effectively as their equivalent welded structures when end-loaded by impact. The joint does not 'unzip' ahead of the plastic deformation of the metal. While this gives an interesting insight into the collapse of a component at high strain rates it can not be assumed that such an observation will be valid for other shapes or similar tubes impacted elsewhere and in a different manner. To avoid beams and joints unzipping in a crash, the extremities of their bonded flanges should be pinned - either mechanically or by a spot weld. This will rarely prove inconvenient since intermittent spot-welding is a useful technique for holding components during the periods when the adhesive is either uncured or in a temporally softened form as a result of a transit through a paint stoving oven. Furthermore, recent work (19) has shown that the impact resistance of a welded aluminium joint is considerably enhanced by the presence of an impact resisting toughened adhesive.

Another facet of impact damage and the question of energy absorption is the relative strain/rate sensitivity of the bonded system as a whole. If energy absorption is a major consideration, then due attention should be paid to both the stiffness of the adherend as well as that of the adhesive and any changes which may occur with the strain rate (18).

Primary Load Bearing Areas

These encompass all the bonded joints which comprise an integrated monocoque body shell together with the bonded hinges and lock mechanisms of secondary structures, particular reference being made to those structures which are involved in either energy absorption or power delivery. Typical examples are suspension related components and bonded joints in drive shafts. While neither may be seen for some time, the former is probably unavoidable! In such situations only proven practice, supported by 'Codes of Practice', should be permissible. At the moment there is little supportive practical experience generally available and so for the time being progress must, of necessity, be slow. Nonetheless, the fact that BLT's ECV-3 exists augers well for the future.

Repair

There is little doubt that the subject of vehicle repair raises major issues. Studies already in hand indicate quite clearly that the future bonded vehicle needs to be designed in order to facilitate repair. Repair techniques and materials need to be developed and the trade itself will need regulation and training. Unless a concerted effort is made immediately by all concerned the bonded vehicle will fail to develop its full potential as soon as it otherwise might.

Adherend Problems

There is a general appreciation that adhesives require a correctly prepared adherend surface if they are to maintain the integrity of a primary structural joint for 15-20 years. However, what is not commonly realised is that some surfaces present the adhesive manufacturer with very difficult problems, a typical example being presented by one of the many types of pre-treated steel. These range from some form of simple zinc plate to complex coated alloys based on aluminium and/or zinc. A finishing coat of paint or some form of plastics films may also be added. Obviously, the latter cannot carry substantial loads and there is a general appreciation of this fact. Regrettably, there is no similar recognition of the potential weakness of some metallic protective coatings. A strong adhesive can readily tear such a coating from the surface of the underlying steel. The effect tends to become even more pronounced as the thickness of the plate, or coating alloy, increases or if the underlying sheet metal is distorted.

Related problems are caused by the presence of weak, friable oxide layers on the surface of a zinc plate. These difficulties, which are very similar to those observed on oxidised aluminium surfaces, will respond well to the appropriate chemical treatment. By contrast, the adhesion of the protective layer of zinc presents an obstacle of greater dimensions. The real answer to this particular problem must lie in the hands of the metallurgist. However, an interim solution is the creation of adhesives whose shear modulus is such that it will distribute the imposed load properly and prevent the anticipated stress approaching the level which will rupture the steel/zinc interface. Such a technique has been successfully employed in conjunction with composites whose surface - whether polyester or epoxy based - always tends to be brittle. The reduction in the shear load at the edge of the joint demonstrated in Fig.2 compared with Fig.3 illustrates this point clearly. Related work by Green & Bowyer (20) on the fabrication of GRP sheds an interesting side light on this issue.

Adhesive Problems

Although toughened adhesives have been available commercially for some 10 years (4,21) and have undoubtedly proved that they are viable candidates for structural assembly, it is a fact that as yet there are relatively few that have been thoroughly studied and which have a proven record. It is equally true that enormous efforts are being made to change this situation by the industry at large - but inevitably, some time must elapse before fully developed groups of related products become freely available to the designer. In the meantime he must proceed with caution.

General Considerations

Computer programmes are now becoming available which will enable designers to examine the interplay between the moduli of the materials used, the geometry of the joint and the various dimensions involved. Figs. 2 and 3 were developed from just such as evaluation.

In the absence of such specialised programmes it is probably safest to assume that for sheet metal approximately 1 mm thick that all the load will be carried within the first 6 mm or so of each end of the overlapped area if a stiff adhesive is used. For this reason, the minimum overlap for any form of load bearing structure should be at least 12 mm. Once the overlap length exceeds 10/12 mm the central area ceases to be stressed to any significant degree. When such an overlap is used in conjunction with a high performance structural adhesive, the joint's performance and ultimate failure will be significantly affected by the modulus of the adherend. It is often difficult, if not impossible, to say which is responsible for the joints ultimate rupture as the load rises to an intolerable extreme.

The stress distribution in a typical lap joint subject to a variety of loads is illustrated in Fig.5. For comparison, the changes in stress generated in five lap joints of differing length is given in Fig.6. The effect of this on the apparent strength of the joint is shown in Fig. 7. The discrepancy - which is commonly misunderstood - is clearly shown.

Figure 5

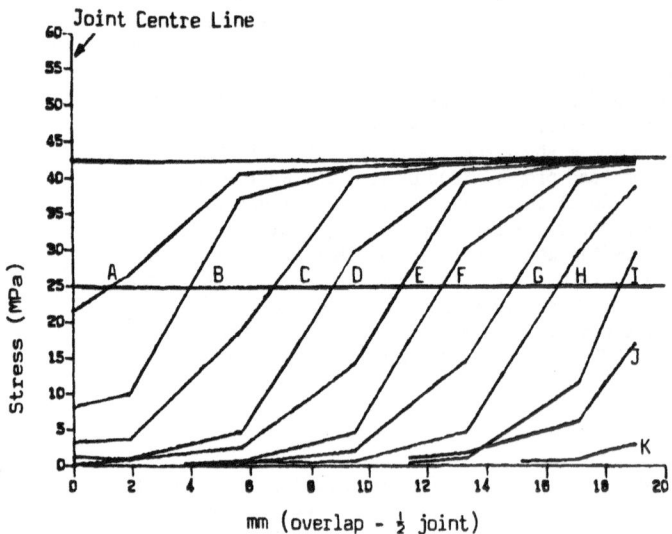

mm (overlap - ½ joint)

Load

A	=	112 KN
B	=	100 KN
C	=	84 KN
D	=	68 KN
E	=	56 KN
F	=	44 KN
G	=	32 KN
H	=	20 KN
I	=	8 KN
J	=	4 KN
K	=	1 KN

The changing pattern of stress distribution displayed by a computer model as the epoxy based steel/steel joint shown in Figure 3 is progressively loaded from 1000 N to 112,000 N.

Note 1: Half joint displayed

Note 2: The asymptotic shear stress and elastic limit are as given in Figure 3.

Note 3: The metal adherends of the joint could be expected to fail in the area of loads E-F. The computer model has been run on to illustrate the progressive load transfer towards the inner areas of the joint as the load is increased. This could be expected with heavy steel plate.

Note 4: The combination of the approach of the adhesive to its asymptotic shear strength combined with the elastic/plastic deformation of the thin sheet steel normally causes cohesive failure of the adhesive. If the distortion is prevented - in thin sheet metal - then the adherend fails.

Note 5: Between loads J and K a creep and fatigue resistant joint is created - the central areas are not loaded and are stable.

Figure 6

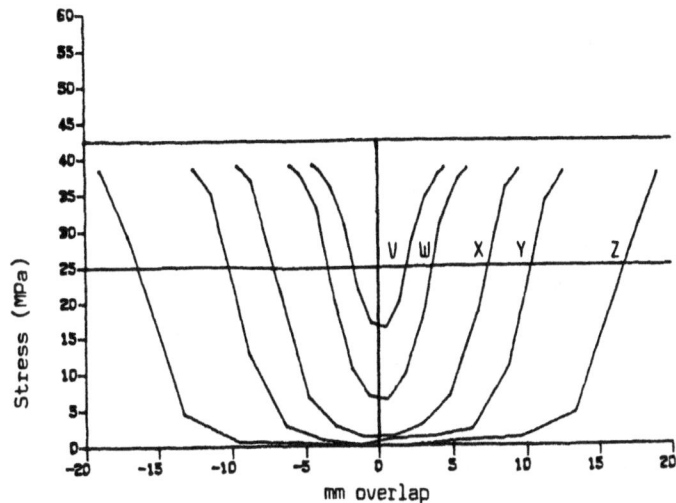

V = 9mm overlap
W = 12mm overlap
X = 19mm overlap
Y = 25mm overlap
Z = 38mm overlap

The changing stress pattern displayed as the length of the overlap
of the joint described in Figure 3 is reduced from 38mm to 9mm
under an imposed load of 20KN.

Note 1: For lengths X, Y and Z the stress pattern overall is
        similar - though obviously the unloaded central area
        is reduced in the joints with the shorter overlap.

Note 2: At 12mm the load must be borne in the central areas.

Note 3: At 6mm, not shown, the joint cannot sustain the load at
        all.

This disproportionality means that beyond a certain point it is not
possible to increase the strength of a joint simply by increasing
the overlapped length. It can only be done by changing other parameters
such as the width of the joint, the thickness or nature of the
adherend and possibly by substituting a more appropriate adhesive.
However, an increased overlap will not only improve the environmental
performance, it will also maximise creep and fatigue resistance.

Figure 7

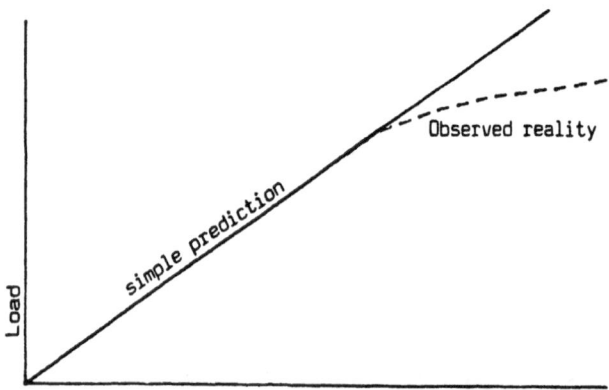

Joint Area (width static - length increasing)

Failure of performance predictions based on simple unit area
extrapolation to match observed performance. This follows from the
unloading of the central joint areas as the length of the joint
increases. See VWX versus XYZ of Figure 6.

The simple assumption which may be made for the thinner sheet metals
may not be valid where thick sections are employed. Many relatively
simple joints which will not be subjected to extremes of load or
environment may be based upon the following empirical rules :-

*   Minimum overlap of 12mm (5-6mm for nominal loads).

*   Maximum overlap of 25mm (this allows for accidental damage
     either within or alongside the joint area).

*   The requisite load bearing area being based on the
     recommended overlap length and the appropriate width.

*   The joint should be designed to minimise peel and cleavage
     forces though the possibility that these may arise as a
     result of accidental damage should be considered.

*   Employ the maximum rigidity or thickness of component
     compatible with other cost/weight considerations.

*   If possible, use contained, or slotted, joints - particularly
     those generated by the profile of an extrusion or pultrusion.

*   Fatigue resistance is maximised by ensuring that tensile
     shear loads do not oscillate between compression and
     tension and that the elastic shear stress limit of the
     adhesive is not exceeded.

It will be noted that no account has been taken of the thickness of the
adhesive within the joint line. Conventional wisdom states that the
bond line should be as thin as possible. This is usually true for
the older types of conventional high strength - though brittle -
adhesives. However, while it is generally true for the new 'Toughened'
non-brittle adhesives - their peel strength normally <u>increases</u> with

the thickness of the adhesive layer. This is very important since, in practice, it is the peel and cleavage forces induced by accidental impact that usually destroy joints. Further work is needed in these areas to quantify the general observations.

Evaluation

Conventional laboratory lap joints can be used to give a realistic assessment of the likely durability of an equivalent section of a bonded components and there is a considerable literature devoted to this subject (22). However, the performance of lap, 'T' peel, impact and other standard laboratory joints <u>cannot</u> be extrapolated and considered to be an indication of full scale performance. Appropriate data may only be obtained from the components themselves which may be progressively extrapolated as confidence in mathematical modelling techniques grows. This knowledge once gained and proven should be built up into a Code of Practice.

Practical Points

Apart from the physical problems of actually being able to assemble components sufficiently well to make an effective joint, there are a number of practical considerations which need to be taken into account. For example, when two components have to be assembled using a lap joint configuration the deliberate act of shaping - perhaps by slight indentation - one part may allow the more accurate placement of the second. Similarly, care must be taken in either the choice of adhesive or in the assembly technique devised to ensure that the adhesive itself is not wiped from the surface as the parts are put together. Placement, not sliding techniques, should be employed for the latter tend to wipe the adhesive from the joint area and may lead to the formation of air pockets and consequently reduce the overall performance. A suitable adhesive selection procedure (23) will eliminate adhesive families and their sub-groups which are unlikely to meet the basic requirements of any particular design. However, the designer should take this process further - preferably with the assistance of the manufacturer of the selected type of adhesive in order to identify or prepare the individual formulations which will most readily meet production requirements. The most suitable adhesives will :-

*   Be safe

*   Store well

*   Not require mixing

*   Dispense readily through simple equipment which will lend itself to automation and associated techniques.

*   Cope well with the inevitable variation in production conditions - particularly during the curing process!

*   Resist, in either the uncured or cured state, the damaging effects of subsequent assembly processes.

Surface Preparation

Quite contrary to popular belief - a belief founded upon the
experience of using traditional adhesives - reliable joints can be
formed on a variety of unprepared surfaces.

The introduction of adhesives capable of accommodating surface
contamination has given a degree of freedom which was previously
unknown. Of course, there can be no denying that the better the
preparation the better the overall performance. But, providing
contamination is not gross, perfectly adequate levels of performance
can normally be obtained from :-

> Anaerobic
>
> Cyanoacrylate
>
> Plastisol
>
> Toughened Acrylic and
>
> Toughened Heat Cured Epoxy Based Adhesive

However, the subject of surface preparation should be treated with
respect and the knowledge that preparation may not be necessary
should be considered as an inducement to explore carefully rather
than a licence to proceed without caution. In general, a level of
surface preparation can be chosen to give :-

* Optimum adhesion with good environmental resistance
* Good adhesion with moderate environmental resistance
* Moderate adhesion with low environmental resistance.

Typically, and for most materials, these levels are obtained from
the following processes :-

* Some form of chemical pre-treatment
* Surface abrasion and degreasing
* Degreasing only - or no treatment at all

As has been pointed out it would be unwise to settle for a
potentially debased performance without considering the overall
situation very carefully. For example, some materials are
notoriously difficult to bond and will perform badly almost no matter
what is done. By contrast, other materials will give a good
reproducible performance with minimal attention to preparation. A
simple summary of what might be anticipated is given in Tables 6
and 7 where the relative merits of a variety of common metals and
plastics are listed.

Preparative techniques

While surface preparation is not always required and some adhesives
will cope extremely well with poor surfaces, simple minimal surface
preparation such as cleaning and degreasing should always be
considered. Components made from materials which have weak or loose
surface layers, or are prone to stress cracking, solvent attack or
water migration require special treatment. Techniques which will

ensure that adhesion can be obtained with even the most difficult of
materials are available and widely published and a simple review (24)
of some general methods suitable for the common materials is freely
available.

As the preparative treatments for aluminium and its alloys are
somewhat complex and expensive, it is interesting to note that simple
techniques employing aqueous systems which require neither extreme
pH's or imposed electric currents are now becoming available. A
review of some recent work has just been published (25). It is
understood (26) that such pre-treatments are also effective on steel
and composite surfaces.

Table 6 - Suitability for Bonded Assembly - Metals

Material	Suitability Rating*	Comment
Aluminium and its alloys	2 - 3	Load bearing but surface must be properly prepared for use in severe environments.
Copper and alloys	3 - 4	Do not load. Careful preparation necessary.
Steel (mild)	1 - 2	Load bearing. Really thorough preparation is not always required.
Steel (HSLA)	1 - 2	Load bearing. But experience limited as yet.
Steel (stainless)	2 - 3	Load bearing. Careful surface preparation may not be necessary.
Steel (precoated PVC or paint)	3 - 4	Cannot be loaded but can bond well.
Zinc passivate	2 - 3	Load with care - correct passivate necessary.
Zinc or zinc plating	2 - 3	Load with care. Preparation necessary.

* Scale (1-4)

1. Good   3. Many problems
2. Problems   4. Not recommended.

Table 7 - Suitability for Bonded Assembly - Plastics

Material	Suitability Rating*	Comment
Plastics		
Fibre reinforced GRP		
thermoplastic	?	Only load lightly.
thermoset	1 - 2	Load bearing.
acrylic faced	2 - 3	Do not load.
wood faced	1 - 2	Can be loaded.
Pultrusions	1 - 2	Load bearing.
SMC	1 - 2	Load bearing.
CFRP	2	Load bearing.
Other Plastics		
ABS	3 - 4	Only load very lightly.
Nylon	2 - 3	Load lightly.
Polyolefin)plus variants)on basic type)all of which)may or many not	4	Do not load.
Poly-)be filled to urethane)some degree	3	Light loads.
PVC (rigid)	2 - 3	Only light loads.

*Scale (1-4)

1.	Good	4.	Not recommended
2.	Problems	?	Not enough known
3.	Many problems		

QUALITY CONTROL

As no totally effective NDT method has yet been devised for the assessment of bonded joints, the process of Quality Assusrance must be carried out indirectly. Table 8 lists the more significant aspects of a Quality monitoring system.

SUMMARY

A bonded aluminium based vehicle has now been built (6) and its existence is a clear demonstration of both the designer's skills and the recent advances in adhesive technology. However, while exciting, this step should only be regarded as an indication of future trends. While all the basic tools and materials - computer programmes, toughened adhesive, assessment and characterisation techniques and simplified preparative methods are now becoming

Table 8 - Quality Assurance - Assessment of Both Adhesive and Bonded Structure

Adhesive Manufacturer

* QA assessment of Raw Material supplier where appropriate.
* QA assessment of manufacturer by qualified external organisation.

Adhesive Production (See Property profile Table 3)

* Manufactured according to agreed Specification.
* Batch acceptance test on agreed characteristics.
* Random type tests on Primary Properties.

Vehicle Manufacturer

On delivery -

* Repeat batch tests unless certification in force.

Periodically -

* Random batch and type tests on stock.
* Check stock rotation.

During production -

* Monitor surface condition of adherend in critical areas.
* Monitor application of adhesive in critical areas.
* Monitor adhesive curing regimes.
* Prepare test coupons in parallel with assembly of components and circulate through the manufacturing processes as appropriate.
* Constantly monitor performance of test coupons - automate this process and monitor cumulative data.
* Random assessment of environmental performance of test coupons and relate to previous data.
* Random destructive assessment of component performance.

Vehicle in use -

* Monitor field performance by assessing joints in scrapped (crashed) cars.

available it will still be some time before the many aspects involved can themselves be successfully assembled by those concerned into a working practical technology - a technology capable of producing a bonded car every minute, day in and day out. A car which will be both cost effective to buy and to maintain over a life span of 15-20 years. The omens are good - could it be that we will eventually see a car which is recycled through a refurbishing plant rather than a scrap heap? In principle, there is no reason why not. A bonded corrosion resistant monocoque shell could be refitted and refurbished.

Would it be economic and perhaps more pertinently - would it be acceptable?

Speculative though this may be there is little doubting the managerial skills which will be needed to make effective progress. These will include - education, inter-industry involvement, commitment to technical evolution, the effective use of all available talent and perhaps above all perseverance.

BIBLIOGRAPHIC REFERENCES

1. Noton, B.R., "Swedish Aeronautical Research" A lecture presented at the Conference on Adhesive Bonded Structures for Aircraft, Los Angeles, Calif; January 31- February 2. 1957.

2. B.F. Goodrich, information brochure, 'Toughen Epoxy Resins with Hycar RLP'.

3. Du Pont, Manufacturer's information, Private Communication, 1975.

4. Lees, W.A., (1981), "Modified Epoxides ; Practical Aspects of Toughening", Journal of Adhesion, Vol.12, pp.233-240.

5. Lees, W.A., (February 1981), "Use of Adhesives in Constructing Vehicles", Adhesives Age, pp.23-31.

6. King, C.S., (1983), "A car for the nineties : BL's energy conservation vehicle", Proceedings Institution Mechanical Engineers, vol.47, no.64.

7. Lees, W.A., (May 1980), "Bonding metal structures with the new toughened adhesives", Sheet Metal Industries, pp.447-462.

8. "Designing for stress in plastics", (September 1983), Engineering Materials and Design, p.53.

9. Sigman, D.R., Buechel, J.H. and Ervin, P.R., (1983), "Evaluation of the Ford GrfRP lightweight car", International Journal of Vehicle Design, vol.4, no.6, pp.557-570.

10. Koht, L. and Hall, T.R., (June 1975), "Soviet Glue-Welding Technology", Article V, Advancements in Materials Science, Report No.1.

11. Kinloch, A.J. and Young, R.J., (1983), Fracture behaviour of polymers, London, Applied Science Publishers Limited.

12. Frazier, T.B. and Lajoie, A.D., "Durability of adhesive bonded joints", Air Force Materials Lab., TR-74-26, Drayton, Ohio, March 1974.

13. Adams, R.D. and Peppiatt, N.A., (1974), "Stress Analysis of Adhesive-bonded Lap Joints", Journal of Strain Analysis, vol.9, no.3, pp.185-196.

14. Beevers, A. and Kho, A.C.P., (January 1983), "The performance of adhesive-bonded thin gauge sheet metal structures with particular reference to box section beams", International Journal of Adhesion and Adhesives, pp.25-29.

15. Mao, T.J., (1983), "Automotive Applications of Plastics", AAAS Symposium - The Chemistry of Transportation Materials, Detroit.

16. Labana, S.S. and Cotton, F.R., (1983), "New uses for adhesives in automobiles", Adhesives Age, May, pp.31-32.

17. "Adhesives Stretch the strength of Box Sections", Review, Eureka, February 1983, pp.48-50.

18. Adams, R.D., Bristol University, Private Communication, November 1982.

19. Welding Institue (1983), Members Confidential Report.

20. Green, A.K. and Bowyer, W.H., (January 1981), "The development of improved attachment methods for stiffening frames on large GRP panels", Composites, pp.49-55.

21. Wake, W.C., (1982), Adhesion and the Formulation of Adhesives (2nd Edition), London, Applied Science Publishers Limited.

22. Kinloch, A.J., (1983), Durability of Structural Adhesives, London, Applied Science Publishers Limited.

23. White, L.M., (1981), Fulmer Materials Optimiser - 2nd Edition, Section I.D.-J.b.5.1, Stoke Poges, Fulmer Research Institute Limited.

24. Engineers Guide to Adhesives, (1983), Eastleigh, Permabond Adhesives Limited.

25. Moulds, R.J., (1984), "Some effects of alloying elements on the performance of bonded aluminium alloy joints", International Journal of Adhesion and Adhesives, January.

26. Permabond Adhesives Limited, (1983), Private Communication.

Chapter 9

MOISTURE-CURED POLYURETHANE SEALANTS MODIFIED
WITH BITUMEN-ISOCYANATE ADDUCTS

Janusz Kozakiewicz and Andrzej Lendzion

Institute of Industrial Chemistry, 01-793 Warsaw, Poland.

INTRODUCTION

One-component moisture-cured elastic polyurethane sealants are
among the premier sealants of the 1980s [1,2] mainly because of their
excellent mechanical properties and weatherability. They are used in
construction, civil engineering and ship of yacht building to fill joints
of concrete, wood, glass, aluminium and plastics and in the machine and
automobile industry to seal metal parts [3-5].

Various systems have been proposed so far to achieve moisture-
curing of polyurethane materials. These include [6]:
- isocyanate-terminated urethane prepolymer which is able to
  react directly with atmospheric moisture [4,7,8]
- blocked urethane pre-polymer mixed with blocked curing agent
  (e.g. ketimine) that is released immediate when brought into
  contact with moisture [9,10].
- alkoxysilane-terminated polyurethane obtained by reacting
  isocyanate-terminated urethane pre-polymer with amino-
  trialkoxysilane. Here, moisture-curing proceeds the same
  way as for R.T.V. silicone rubber [11].

The system based on an isocyanate-terminated pre-polymer is obviously the simplest and therefore the most commonly used one. The moisture-curing of such systems is represented in Fig. 1.

MOISTURE CURING OF POLYURETHANE SEALANTS

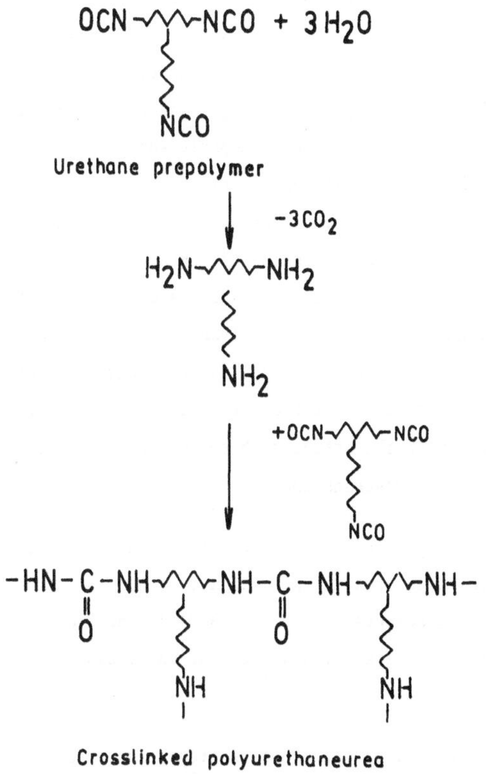

$$OCN \text{-}\!\!\sim\!\!\!\wedge\!\!\!\sim\text{-} NCO + 3 H_2O$$

$$NCO$$

Urethane prepolymer

$$\downarrow \quad -3CO_2$$

$$H_2N \text{-}\!\!\sim\!\!\!\wedge\!\!\!\wedge\!\!\!\sim\text{-} NH_2$$

$$NH_2$$

$$\downarrow \quad +OCN\text{-}\!\!\sim\!\!\!\wedge\!\!\!\sim\text{-} NCO$$

$$NCO$$

$$-HN-\underset{\underset{O}{\|}}{C}-NH\text{-}\!\!\sim\!\!\!\wedge\!\!\!\sim\text{-} NH-\underset{\underset{O}{\|}}{C}-NH\text{-}\!\!\sim\!\!\!\wedge\!\!\!\sim\text{-} NH-$$

$$NH \qquad NH$$

Crosslinked polyurethaneurea

Fig. 1   Moisture curing of polyurethane sealants.

In commercial formulations, isocyanate-terminated pre-polymer, that is the substantial part of the product, is usually mixed with mineral or carbon fillers, thixotropic agents, plasticizers, resins, etc. Especially interesting for the sealant formulator is the possibility of polyurethane modification with bitumens since in this way some new features of the product can be achieved and its cost can be lowered significantly.

However, in spite of this, only a limited number of papers and patents has been published dealing with polyurethane-bitumen blends, especially because of the low compatibility of both substances. Blends containing polyurethanes and coal-tar or coal-pitch, especially when used as solvent-based lacquers and coatings, have been investigated more closely [12-23] but only a little is known of the properties of polyurethane-asphalt blends. With these later, the poor compatibility of the components has been overcome either by high temperature blending of urethane pre-polymers with melted asphalt [24,25] or by reacting substrates for polyurethane directly in melted asphalt [26-29]. These methods have been successful only for limited proportions of asphalt and polyurethane components. An attempt at improving their compatibility by adding carboxylic acids has been reported [30], but better results are obtained when asphalt-isocyanate adducts are used instead of asphalt [31,32]. Reacting active, hydrogen-containing asphalt ingredients with isocyanate before blending with isocyanate-terminated pre-polymer improved compatibility and makes impossible undesirable reactions during blending. This method was applied also in the studies carried out at the Institute of Industrial Chemistry in Warsaw on the effect of bitumen addition on the properties of moisture-cured polyurethane sealants.

It was found earlier in preliminary investigations [33,34] that blending of isocyanate-terminated urethane pre-polymer containing carbon filler with asphalt-isocyanate adducts of various -NCO content led to the products which could be useful as special sealants for the automobile industry.

In this paper, the rheological and adhesion properties of such blends will be discussed and the results of their testing as sealants for the automobile industry will be presented.

As urethane components, coal dust/carbon black filled polyurethane moisture-cured sealants were applied. They were prepared according to procedures previously described [35,36] from tolylene di-isocyanate and either polyoxypropylene trio of molecular weight 3000 (PU-1) or polyoxypropylene diol of molecular weight 2000 (PU-2) using a NCO/OH ratio ca. 1.6/1.0. Asphalts used in these studies were two petroleum asphalts of m.p. 80°C: block (A-1) and not-blown (A-2). Asphalt-isocyanate adducts (AA-1 and AA-2) were prepared by reacting 100 parts of asphalt with 7 parts of tolylene diisocyanate in 20 parts of toluene at 90°C. PU/AA blends were obtained by kneading both components in a laboratory mixer over 30 min. Products obtained in this way were black-coloured viscous pastes. After 14 days of moisture-curing at ambient conditions (25°C, 50% R.H.) they formed elastic, elasto-plastic, or plastic materials depending on the PU/ AA ratio. As it can be seen in Figure 2, their hardness, as determined by needle-penetration measurements, clearly decreased with increasing AA content.

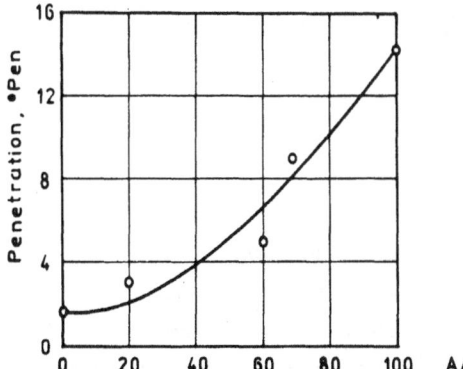

Fig. 2   Needle penetration of cured polyurethane sealant modified with asphalt-isocyanate adducts vs. AA content in PU/AA blend. PU-1 and AA-1 were used.

RHEOLOGICAL STUDIES

To examine rheological properties of PU/AA blends, samples were sheared in a Rotovisko Viscometer for 5 min. at a rate of 44.1 s^{-1} and then for 10 min at a rate of 4.9 s^{-1}.

Results of rheological studies are presented in Figures 3-5. It can be clearly seen in Figure 3 that PU sealant became thioxotropic when blended with asphalt-isocyanate adduct AA-1. While the viscosity of PU alone was more or less constant with shear rate and shearing time, a distinct dependence of viscosity on both shearing time and shear rate appeared for the PU/AA blend (compare lines 2 and 3 in Fig. 3). The viscosity of this blend measured at high shear rate rapidly decreased with shearing time and when the shear rate was switched to low value it immediately rose to almost 5 times greater level.

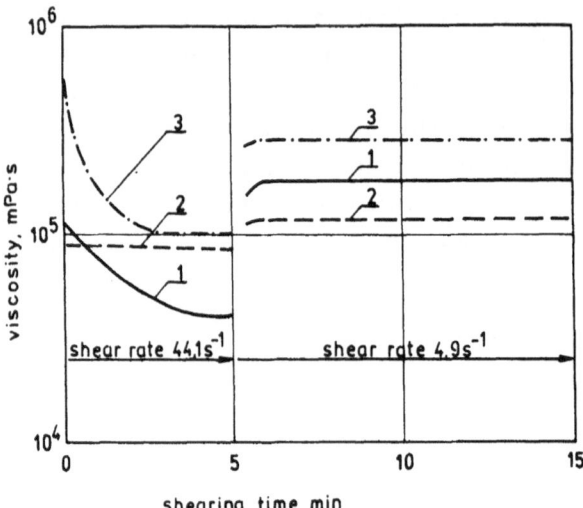

Fig. 3   Viscosity of:

asphalt-isocyanate adduct (AA) – curve 1,

polyurethane sealant (PU)   – curve 2

PU/AA = 40/60 blend   – curve 3

Measured for low and high shear rate vs. shearing time.

PU-1 and AA-1 were used.

This phenomenon is very important from a practical point of view
since the application requirements for sealants include both good gunna-
bility (which means: low viscosity at high shear rate) and lack of flow
on vertical surfaces (which means: high viscosity at low shear rate)[37,38].

The effect of AA content in the blend with PU on high shear rate
viscosity η_1, measured after 5 min shearing at a rate of 44.1 s^{-1} and
on low shear rate viscosity η_2, measured after 1 min shearing at a rate
of 4.9 s^{-1}, is presented in Figure 4. It can be seen in Figure 4 that
both viscosities attained a maximum for ca. 60% AA in the blend. It is
interesting to know how the η_2/η_1 ratio (which can be considered as a
measure of non-sagging properties) depends upon the AA content in the
blend. The result is presented in Figure 5. It is shown there that the
best non-sagging properties can be obtained when the AA content is close
to 40%. This was proved by direct measurements of flow on vertical
surfaces made for PU/AA blends. The best results were observed when
samples contained 20-60% AA which corresponds to log $\eta_1/\eta_2 \geq 0.45$.

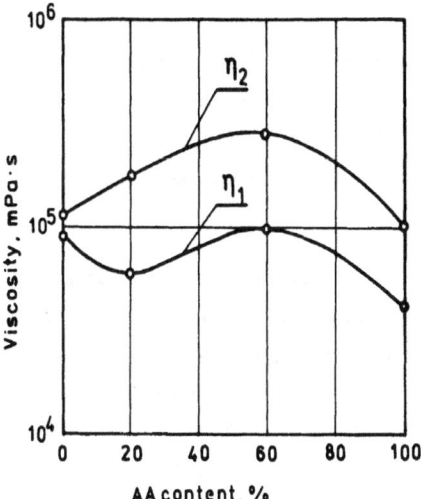

Fig. 4   High shear rate η_1 and low shear rate η_2 viscosity
         values of polyurethane sealant modified with asphalt-
         isocyanate adduct vs. AA content in PU/AA blend.
         Pu-1 and AA-1 were used.

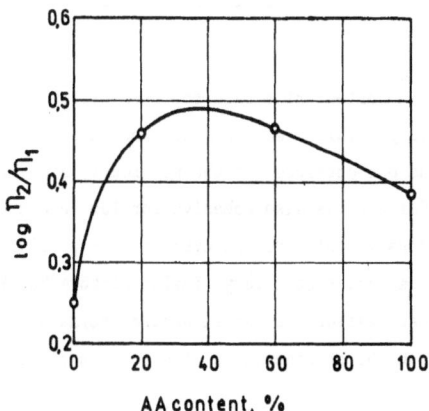

Fig. 5   Thixotropic properties of polyurethane sealant
         modified with asphalt-isocyanate adduct vs. AA
         content in PU/AA blend.
         PU-1 and AA-1 were used.

ADHESION STUDIES

To examine adhesion properties of PU/AA blends, shear strength was measured with standard 10 x 2 cm plain steel plates (bond area: 2.5 cm^2) solvent degreased, after 14 days curing. The results are presented in Figures 6-10 where shear strength is plotted vs. AA content in the blend with PU sealant.

As is shown in Figure 6, interesting behaviour of shear strength was observed as AA content was increased. At first - up to ca. 20% AA - it slowly decreased and then rapidly dropped down to almost six times lower a value while AA content was increased only threefold. After that - over ca. 50-60% AA - again only a slow decrease in shear strength with increase in AA content was observed.

The mode of bond failure also changed with AA content. The failure was purely adhesive when AA content in the blend was low. It became partially cohesive between 15 and 50% AA and then purely cohesive.

Since the phenomenon of change of characteristic adhesion proper- ties of AA-modified PU sealant with increasing AA content seemed discus- sion-provoking, more detailed examination of it was made using another PU/AA system (PU-2, AA-2).

The shear strength - AA content plot made for this system is shown in Figure 7. Though the strength values were higher, the shape of the curve was very similar to the one obtained for the previously studied system (PU-1, AA-1). Bond failure was also adhesive for low levels of AA in the blend, then became adhesive-cohesive and eventually - purely cohe- sive. Taking into account the fact that a very similar picture had been found during preliminary investigations for still another PU/AA system [33], it can be concluded that the phenomenon is generally characteristic of PU/AA blends.

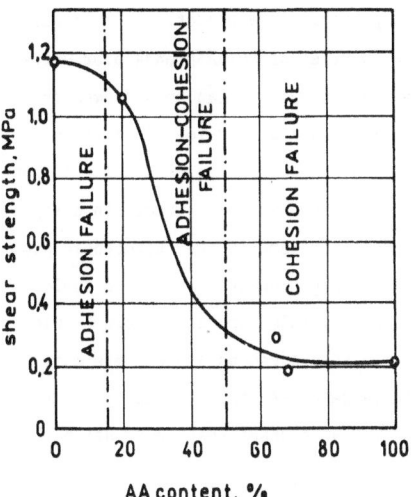

Fig. 6   Shear strength of polyurethane sealant modified with asphalt-isocyanate adduct vs. AA content in PU/AA blend. PU-1 and AA-1 were used.

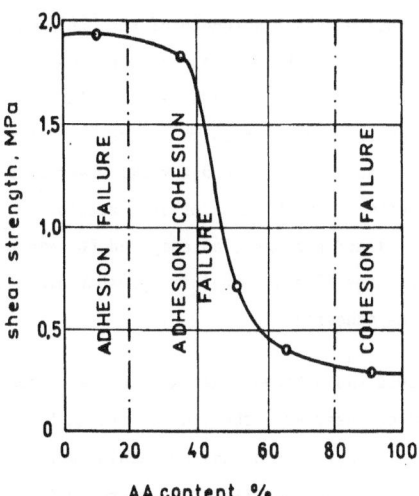

Fig. 7   Shear strength of polyurethane sealant modified with asphalt-isocyanate adduct vs. AA content in PU/AA blend. PU-2 and AA-2 were used.

It seems that the most probable interpretation of the phenomenon of rapid change in adhesion properties of PU/AA blends over certain ranges of AA content is phase inversion. It can be presumed that up to a certain AA content the PU/AA blend is a dispersion of AA in PU where PU is dispersing and AA dispersed phase. When this particular AA content is exceeded, phase inversion occurs which manifests itself in a rapid decrease in shear strength and improvement in the thixotropic properties of the blend. The better non-sagging properties observed inside the region of AA content where phase inversion proceeds would be natural since it is well known that more heterogeneous systems are more thixotropic [39]. Over a certain value of AA content, the system becomes a dispersion of PU and AA which leads to stabilisation of strength and decrease in thixotropy.

Structural studies of PU/AA blends are required to prove this assumption but bitumen/polymer blends are quite difficult to examine under the microscope [40]. However, such investigations have already been started in this Institute.

The distinctive behaviour of shear strength of the bonds made using PU/AA blends of increasing AA content, did not change when the test samples were immersed for over 24 hrs in H_2O, 20% NaCl and fuel oil (Figures 8-10). Only the transition from adhesive to cohesive failure proceeded more rapidly. While adhesive failure was still observed for 35% AA, it became purely cohesive for only 50% AA. After immersion in water or 20% NaCl (Figures 8 and 9) shear strength, as compared to samples not immersed, was only lowered for the blends containing greater amounts of PU, i.e. where the failure was adhesion.

Quite a different result was obtained when samples were immersed in fuel oil (Figure 10). Here, shear strength did not change for blends cntaining more PU but dropped to zero for blends containing more than 60% AA. This feature can be useful in some applications where PU/AA blends are used as anticorrosion coatings since here the essential requirement is that they should be easily removable with gasoline, fuel oil or hydrocarbon solvents.

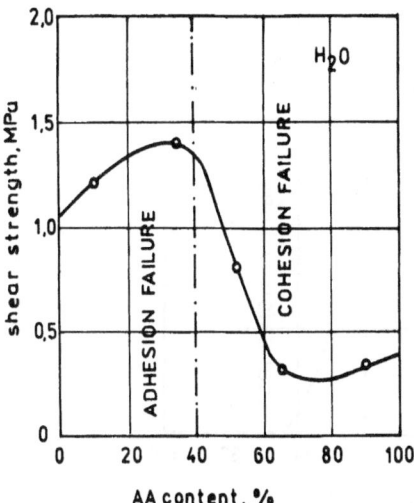

Fig. 8   Shear strength of polyurethane sealant PU modified
with aphalt-isocyanate adduct AA after immersion in
H_2O vs. AA content in PU/AA blend. PU-2 and AA-2
were used.

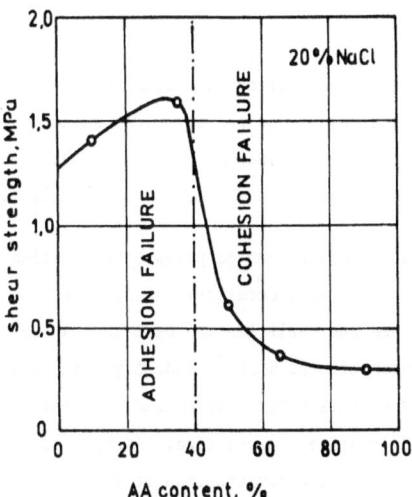

Fig. 9   Shear strength of polyurethane sealant PU modified
with asphalt-isocyanate adduct AA after immersion
in 20% NaCl vs AA content in PU/AA blend. PU-2 and
AA-2 were used.

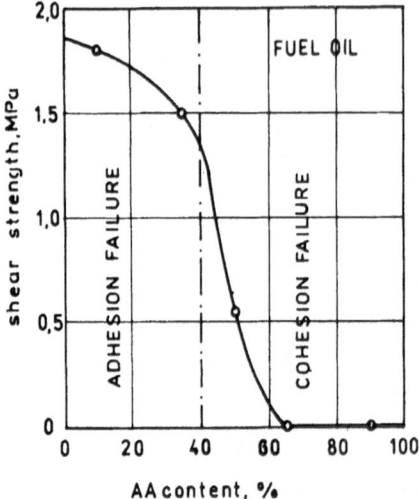

Fig. 10   Shear strength of polyurethane sealant PU modified
with asphalt-isocyanate adduct AA after immersion
in fuel oil vs. AA content in PU/AA blend. PU-2 and
AA-2 were used.

Application tests

Various application tests were carried out using the PU/AA blends
investigated in this study. The results of testing PU/AA blends of relatively high PU content as moisture-cured sealant for the automobile or
machine industry will be presented here. They are shown in Table 1.

These results show that a moisture-cured polyurethane sealant
developed in the Institute of Industrial Chemistry modified with an
asphalt-isocyanate adduct passed the qualification test and can be used
as a metal sealant for the automobile or machine industry. There are
several possible applications of this particular sealant in that area and
some of them have been already realised in practice. Among others, these
include: sealing of steel sheets in the construction of truck cabs and
sealing of back axles, also in trucks.

PU/AA blends of high AA content are easily soluble in gasoline and
have been used successfully in the form of gasoline solutions as anticorrosion coatings in the automobile industry. The results of studies on
the properties of PU/AA/gasoline mixtures will be published soon [41].

TABLE 1

Results of testing PU/AA blend of high PU content as sealant
for automobile industry

Test	Result
A. Uncured product	
1. Density at 25°C	1.1 g/cm^3
2. Cone-penetration at 25°C	245° Pen
3. Curing time initial full	24 hrs 14 days
4. Non-sagging properties after 24 hrs, 2 mm thick layer at R.T. at 140 + 3°C	no flow no flow
5. Effect of product on lacquer. 2 mm thick layer of product was covered with lacquering primer, dried at 140°C over 30 min, covered with automobile lacquer and dried again at the same conditions	No change in lacquer coating
B. Moisture-cured product	
1. Tensile strength	3.4 N/mm^2
2. Elongation at break	137%
3. Low-temperature flexibility (35°C, 24 hrs, 1 mm thick layer on steel plate)	No breaks when rapidly bent at −35°C to 110° angle
4. Corrosion resistivity according to FIAT Standard Procedure No. 50180 (1500 hrs in salt-spray chamber)	No change on coated surface of steel plate

REFERENCES

1. Klosovsky J.M., Adhes. Age 24(11), 32 (1981).
2. Adhäsion 25, 446 (1981).
3. Damusis A., Sealants, Reinhold Publ. Co., 1967.
4. Bylsma, H.R., Adhes. Age 13(2), 25 (1970).
5. Lucke, H., Adhäsion 14, 364 (1970).
6. Prane, J.W., Polym. News 4(1), 37 (1970).
7. Dearlowe, T.J. and Campbell, G.A., J. Appl. Polymer Sci. 21, 1499 (1977).
8. Jap. Plast. No.7, 31 (1970).
9. U.S. Pat. 3420 800.
10. German Pat. 2 725 589.
11. U.S. Pat. 3632 557,
12. Blomeyer F., J. Oil Colour Chem. Assoc. 55, 977 (1972).
13. Jap. Pat. 71 30 308.
14. Jap. Pat. 72 13 981.
15. Brit. Pat. 1 323 884.
16. Jap. Pat. 73 12 877.
17. Jap. Pat. 73 33 259.
18. Jap. Pat. 73 33 971.
19. Jap. Pat. 73 29 533.
20. German Pat. 2 248 581.
21. Jap. Pat. 74 52 294.
22. German Pat. 1 044 323.
23. U.S. Pat. 3 980 597.
24. U.S. Pat. 3 179 610.
25. Jap. pat. 75 36 524.
26. Polish Pat. Appl. P-208 724.
27. French Pat. 1 345 810.
28. U.S. Pat. 3 738 807.
29. Jap. Pat. 76 57 799.
30. U.S.S.R. Pat. 587 725.
31. Polish Pat. Appl. 237 961.
32. U.S. Pat. 3 372 083.
33. Pogorzelska, Z., Igielska, B., Kozakiewicz, J., Lendzion, A., Ochrona przed korozja 20, 78 (1977).
34. Pogorzelska, Z., Igielska, B., Kozakiewicz, J., Lendzion, A., Proceedings of "INTERANTIKOR 77" Conference, Bratislava (1977).
35. Polish Pat. 120 779.
36. Polish Pat. Appl. P-228 811.
37. Kurtz, K., Adhes. Age 26(6), 21 (1983).
38. Tavakoli, F., Choung, H., Ibid 26(6), 13 (1983).
39. Orchon, S., Ibid 15(7), 21 (1972).
40. Bukowski, A., Piotrowska, K., Polimery - Warsaw, 23, 258 (1978).
41. Kozakiewicz, J. et al., Farbe u. Lack (in the press).

Chapter 10

INELASTIC ELECTRON TUNNELLING SPECTROSCOPY AND THE ADHESIVE INTERFACE

J. COMYN

School of Chemistry, Leicester Polytechnic.

1 PRINCIPLES OF INELASTIC ELECTRON TUNNELLING SPECTROSCOPY

Inelastic electron tunnelling spectroscopy (IETS) was discovered in 1966 by Jaklevic and Lambe,[1] and it records the vibrational spectrum of organic materials adsorbed on the insulator of a metal-insulator-metal junction. These junctions are most frequently of the form of aluminium-aluminium oxide-organic material-lead sandwiches. Such a junction is represented in Fig. 1a.

When a bias, V, is applied across a junction, the Fermi levels become separated in energy as shown in Fig. 1b. Electrons may now tunnel elastically from the filled conduction states in metal 1 to adjacent empty states in metal 2. To a first approximation this elastic current increases linearly with applied bias.

When a monolayer of organic material is adsorbed on the insulator, a tunnelling electron may excite a vibrational mode of the adsorbate and lose energy in the process. Such an inelastic process, shown in Fig. 1c, will require a minimum bias, V_{min}, given by eq. 1.

$$eV_{min} = h\nu \qquad (1)$$

Here e is the electronic charge, h is Planck's constant and ν is the vibrational frequency of the excited mode. Inelastic processes are relatively weak with typically less than one per cent of the electrons tunnelling inelastically. However, they may be distinguished from the elastic background by examining the second derivative d^2I/dV^2. The current-voltage characteristics are shown in Fig. 2. The second derivative is usually measured by superimposing small modulation voltages on a slowly increasing juncting bias and measuring the level of the second harmonic generated by the junction non-linearity.

Hansma has published a review of IETS[2], and also edited a book on the subject[3].

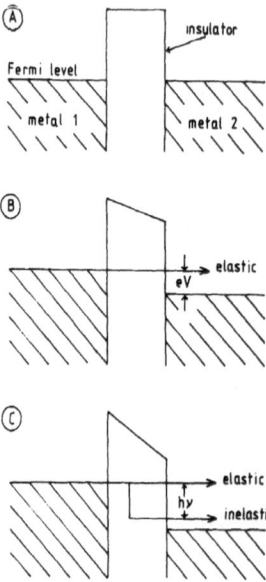

Figure 1 — Elastic and Inelastic Tunnelling at Absolute Zero for Non-Superconducting Metals.

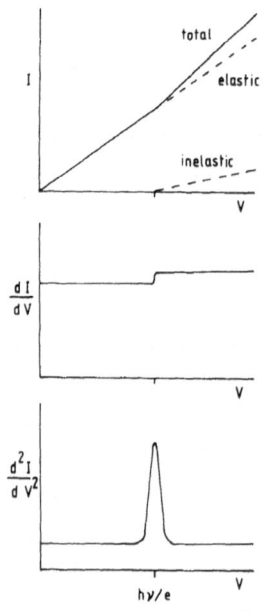

Figure 2 — The Detection of Inelastic Electron Tunnelling.

2   EXPERIMENTAL PROCEDURE

A full description of the spectrometer and the experimental details of tunnel junction fabrication have been published elsewhere[4]. Briefly junctions are prepared on glass microscope slides as shown in Fig. 3. The metal electrodes are deposited by vacuum evaporation, their geometry being controlled by masks. Evaporation of the aluminium electrodes is followed by a glow discharge or venting to the atmosphere which produces an oxide layer approximately 3 nm thick. The slide is then covered with a dilute solution of the organic material and the excess spun off. This method, which is known as liquid phase doping was used with epoxides and amines . But with volatile substances vapour phase doping and infusion doping into a completed junction are also available. In either case the aim is to produce a thin layer of organic dopant roughly a monolayer in thickness. Junctions are completed by evaporation of the lead electrodes; lead is used because it is superconducting at the temperature of liquid helium and so offers improved resolution. Aluminium has been widely studied as a substrate for adhesive bonding; it therefore seemed a possibly fruitful activity to examine some adhesives and adhesion promoters by IETS.

The spectrometer is housed in a room totally screened with aluminium sheets and containing a filtered mains supply. This substantially eliminates both radiated and mains-borne noise so that a high signal to noise ratio is achieved. A block diagram of the spectrometer is shown in Fig. 4.

The small non-linearities in the junction current-voltage characteristics are measured by superimposing a 50 kHz a.c. modulation voltage (usually $\leqslant$ 4 mV peak-to-peak) on a slowly increasing junction bias voltage. The modulation current produces a second harmonic voltage across the junction which is proportional to the second derivative, d^2V/dI^2. This small second harmonic voltage is recovered by a lock-in amplifier and plotted as a d.c. signal on the y-axis of an x-y recorder. The bias voltage is recorded on the x-axis.

3   EPOXIDES AND AMINES

The components studied were the diglycidyl ether of bisphenol A (DGEBA) and two aliphatic amine hardeners, namely di(1-aminopropyl-3-ethoxy) ether (DAPEE) and triethylene tetramine (TETA)[5]. All were soluble in benzene (spectroscopy grade) which did not leave a detactable residue in the junction after doping. The IET spectra shown in Figs. 5, 6 and 7 were obtained from solutions containing 0.05 mg cm^{-3} DGEBA, 0.5 mg cm^{-3} DAPEE and 0.5 mg cm^{-3} TETA respectively.

There is no complete theory of selection rules in IETS. However, there is limited empirical evidence[18] to support the view that where the bonds are aligned

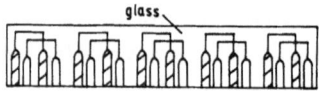

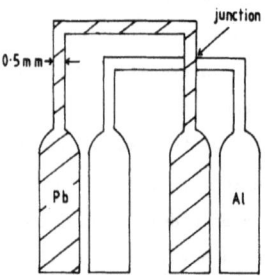

Figure 3   -   A Completed Device Consisting of Five Tunnel Junction and a
Single Junction Shown in More Detail.

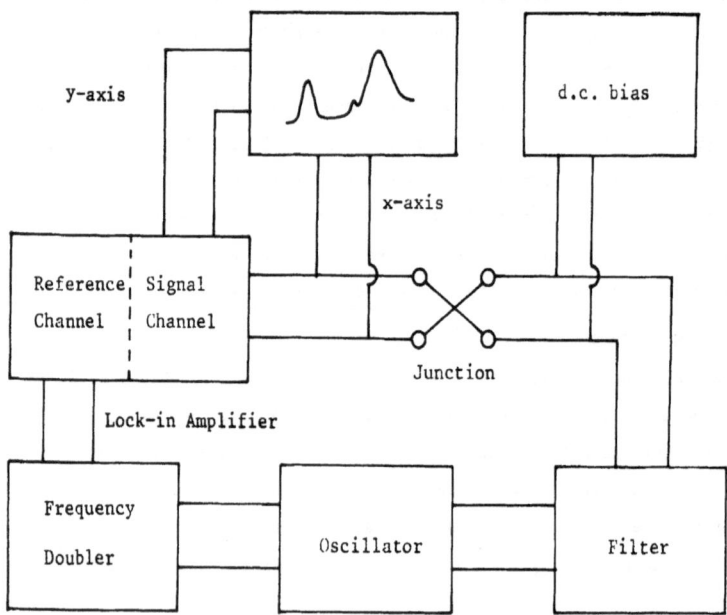

Figure 4   -   Block Diagram of IET Spectrometer.

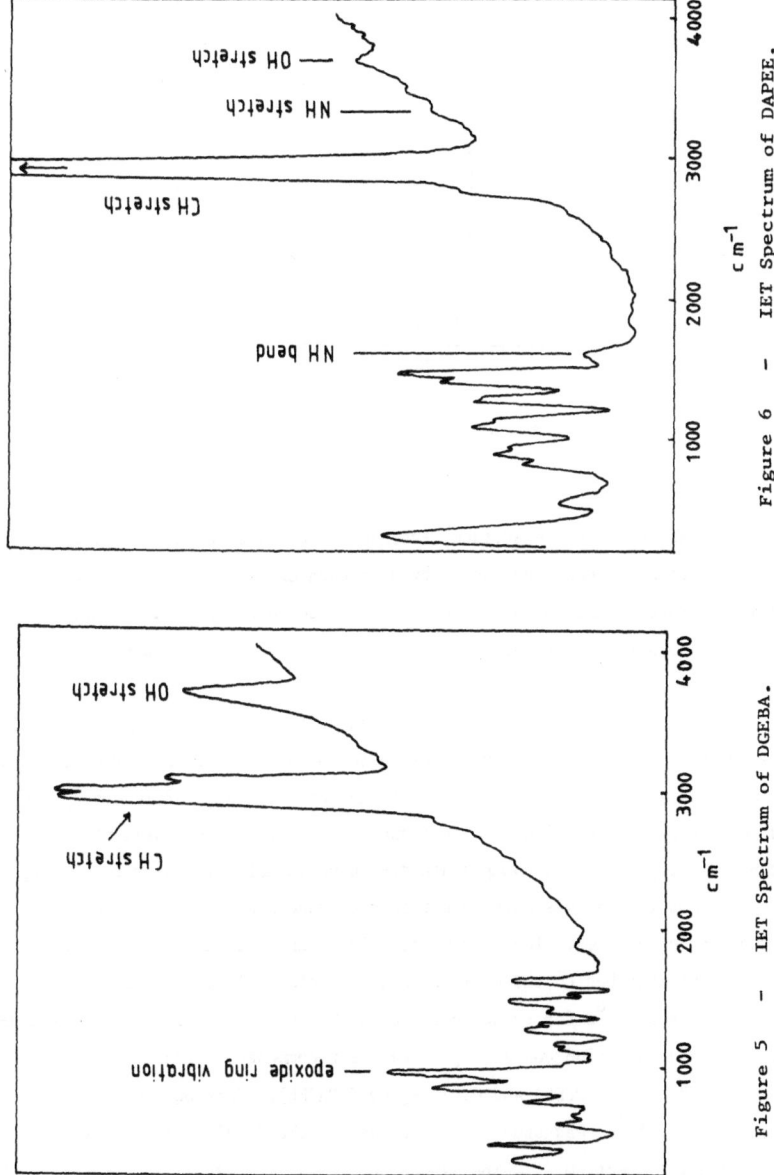

Figure 5 — IET Spectrum of DGEBA.

Figure 6 — IET Spectrum of DAPEE.

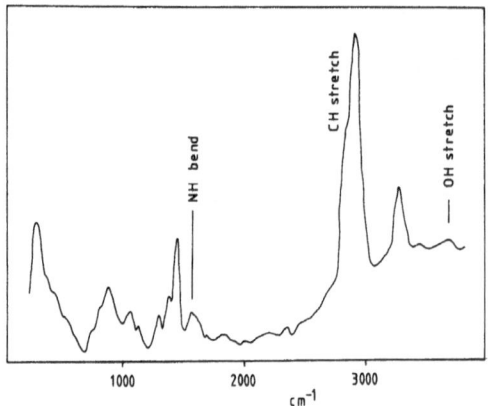

Figure 7 — IET Spectrum of TETA.

perpendicular to the oxide surface, enhanced interaction occurs to give a relatively intense peak. Because they have been exposed to air all the junctions contain hydrated aluminium oxide which gives rise to an –OH peak at about 3630 cm. Another feature which is due to the substrate is the Al–O overtone at about 1600 – 1900 cm^{-1}.

Three curing schedules were investigated for the DGEBA/DAPEE adhesive. These were room temperature curing for up to 70 hours, 80°C for three hours, and 120°C for three hours. To prevent contamination room temperature curing was carried out on completed tunnel junctions in a sealed glass vessel which had been evacuated using a sorption pump containing type 4A molecular sieves. Curing at elevated temperatures was carried out in the evaporator, prior to the deposition of the lead electrodes. The vacuum was $\leqslant 10^{-4}$ τ and heating was by means of a coil (200 W) attached to the masks holding the glass slide in position.

Curing at 80°C for three hours was also investigated for junctions doped with a benzene solution containing twice the recommended concentration of DAPEE (0.033 mg cm^{-3} DGEBA and 0.022 mg cm^{-3} DAPEE). Only one curing schedule, three hours at 60°C, was studied in the case of the DGEBA/TETA adhesive.

The IET spectrum of DGEBA agrees well with i.r. and Raman data, most of the peak positions are within ± 10 cm^{-1} of the corresponding i.r. peaks and ± 20 cm^{-1} of the corresponding Raman peaks. The asymmetric epoxide ring vibration occurs as a strong peak at 939 cm^{-1} in the IET spectrum, compared with a medium intensity peak at 925 cm^{-1} in the i.r. spectrum and a weak peak at

930 cm^{-1} in the Raman spectrum. There is no evidence from the IET spectrum to suggest that DGEBA is chemisorbed onto the aluminium oxide surface, therefore it is assumed that the molecule is attached by physical adsorption.

The IET spectrum of DAPEE is also in close agreement with the i.r. and Raman spectra. The IET spectrum however, in common with that of TETA, shows only weak N-H bending and stretching modes. This suggests that some form of chemisorption of amine groups on the aluminium oxide surface may be occurring.

The IET spectra of the uncured adhesive mixtures show features of the corresponding component spectra. In the spectrum of the DGEBA/DAPEE mixture, the epoxide peak occurs at 932 cm^{-1} and an N-H stretching mode is present at 3320 cm^{-1}; in the DGEBA/TETA spectrum, the epoxide ring absorbs at 930 cm^{-1}.

For both the adhesives and all the curing schedules investigated, the spectra obtained after curing were essentially the same as the spectra of the uncured adhesive mixtures. Also, no significant changes could be detected after curing for the DGEBA/DAPEE adhesive containing excess hardener. The main difference to be expected between the uncured and cured spectra is the disappearance of the epoxide ring due to its reaction with amine.

This reaction can easily be followed using i.r. spectroscopy and by using this technique[6] it has been shown that the curing schedules used are more than adequate to bring about the normal chemical cure of epoxides based on DGEBA and aliphatic amines.

The persistence of the epoxide peak throughout the curing treatments leads to the conclusion that the epoxide ring in DGEBA does not react with amine hardeners within a tunnel junction. This could be due to the epoxide and/or amine groups being adsorbed at the aluminium oxide surface and so prevented from reacting with each other. There is some evidence in the literature to suggest that this may be the case[5].

4 SILANE COUPLING AGENTS

Silane coupling agents can be used as adhesion promoters for glass and metal surfaces prior to adhesive bonding, and the advantage of doing so is that in some instances there is an improvement in environmental resistance. A number of spectroscopic techniques have already been employed to study silanes applied to glass and metal substrates. Infra-red spectroscopy has been used by Boerio and his co-workers to examine 3-aminopropyltriethoxysilane on aluminium[7,8], iron[8,9], copper[10] and titanium[11]. Sung and Sung and their collaborators have examined this same silane upon aluminium oxide using i.r. spectroscopy[12,13] and electron spectroscopy for chemical analysis[13] (ESCA), and have also

demonstrated[14] that it improves the peel strength of aluminium oxide to polyethylene joints. Gettings and Kinloch[15] showed that the silane priming of steel
could improve the environment resistance of adhesive joints, and that this
improvement was directly related to the detection of the Fe-O-Si$^+$ ion in a
static secondary ion mass spectrum.

Solid state carbon 13 nuclear magnetic resonance (n.m.r.)[16] and Raman
Spectroscopy[17] have been used to study silanes on glass, and Bascom[18] has used
i.r., ellipsometry and scanning electron microscopy to examine some silanes
deposited on glass and metal surfaces.

IET spectra of a number of trimethoxy- and triethoxy silanes have been
studied[19], using vapour-phase doping. Atmospheric moisture was not excluded
from the doping vessel as hydrolysis of alkoxy groups is an essential step in
the process of polymerisation of silanes which may occur upon substrates.

4.1 Triethoxysilanes

IET and i.r. spectra of some triethoxysilanes appear in Figs. 8 and 9.
These have six common peaks due to the $Si(OC_2H_5)_3$ group. They are at 1060-1085
cm^{-1} due to Si-O-C, 1088-1114 cm^{-1} due to -O-C-, 1157-1179 cm^{-1} due to Si-O-CH_2,
1273-1310 cm^{-1} due to -C-H bend, and 1370-1403 cm^{-1} due to a CH_3/CH_2 vibrations.
A further peak at 1448-1464 cm^{-1} is due to an asymmetric CH_3 deformation or a
CH_2 deformation. The CH stretching modes at 2900-2980 cm^{-1} are a similar but
less pronounced feature. The additional features are attributable to the
functional groups.

In triethoxysilane (Fig. 8) the functional group is simply hydrogen but
this produces perhaps the most striking spectra with two intense peaks at
880 cm^{-1} and 2191 cm^{-1} which are due to the Si-H bending and stretching modes
respectively.

The vinyl functional group shows a distinct >C=C< stretch at 1597 cm^{-1},
and the peaks at 1011 cm^{-1} and 3047 cm^{-1} are due to the trans CH wagging
vibration and the vinyl CH stretch respectively. In the IET spectrum the SiC
peak at 803 cm^{-1} is weak for vinyltrimethoxysilane, whilst it is strong in the
i.r. spectrum of this compound and also in both spectra for methyltriethoxysilane. It is thought that this is due to interaction between tunnelling electrons
and the π-bond. It will be seen later that the situation is similar for
vinyltrimethoxysilane.

3-aminopropyltriethoxysilane gives a well defined IET spectrum (Fig. 9)
which is very similar to methyltriethoxysilane. The obvious differences are
the NH stretch at 3255 cm^{-1} and the NH deformation at 1590 cm^{-1}. The broad weak

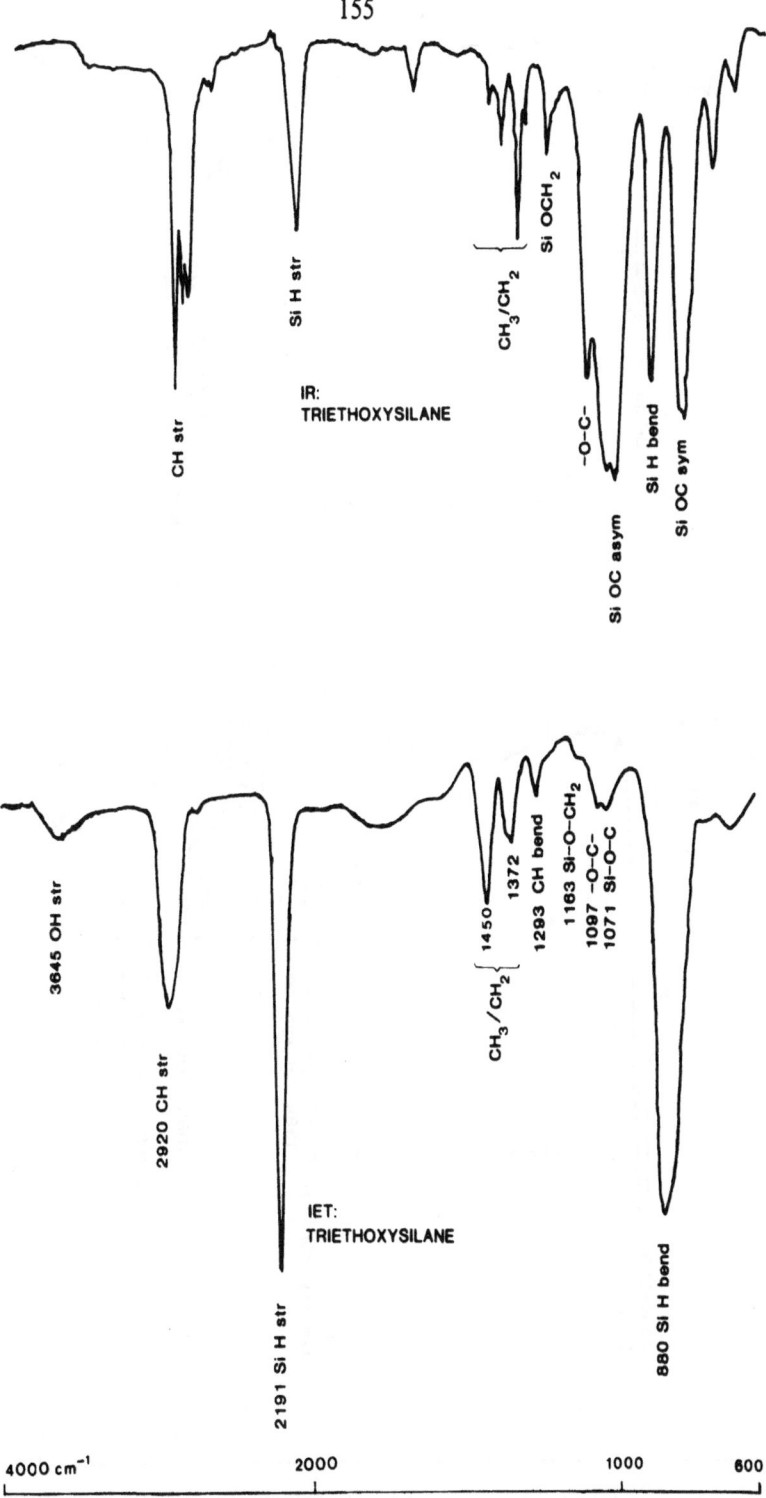

Figure 8 — IR and IET Spectra for Triethoxysilane.

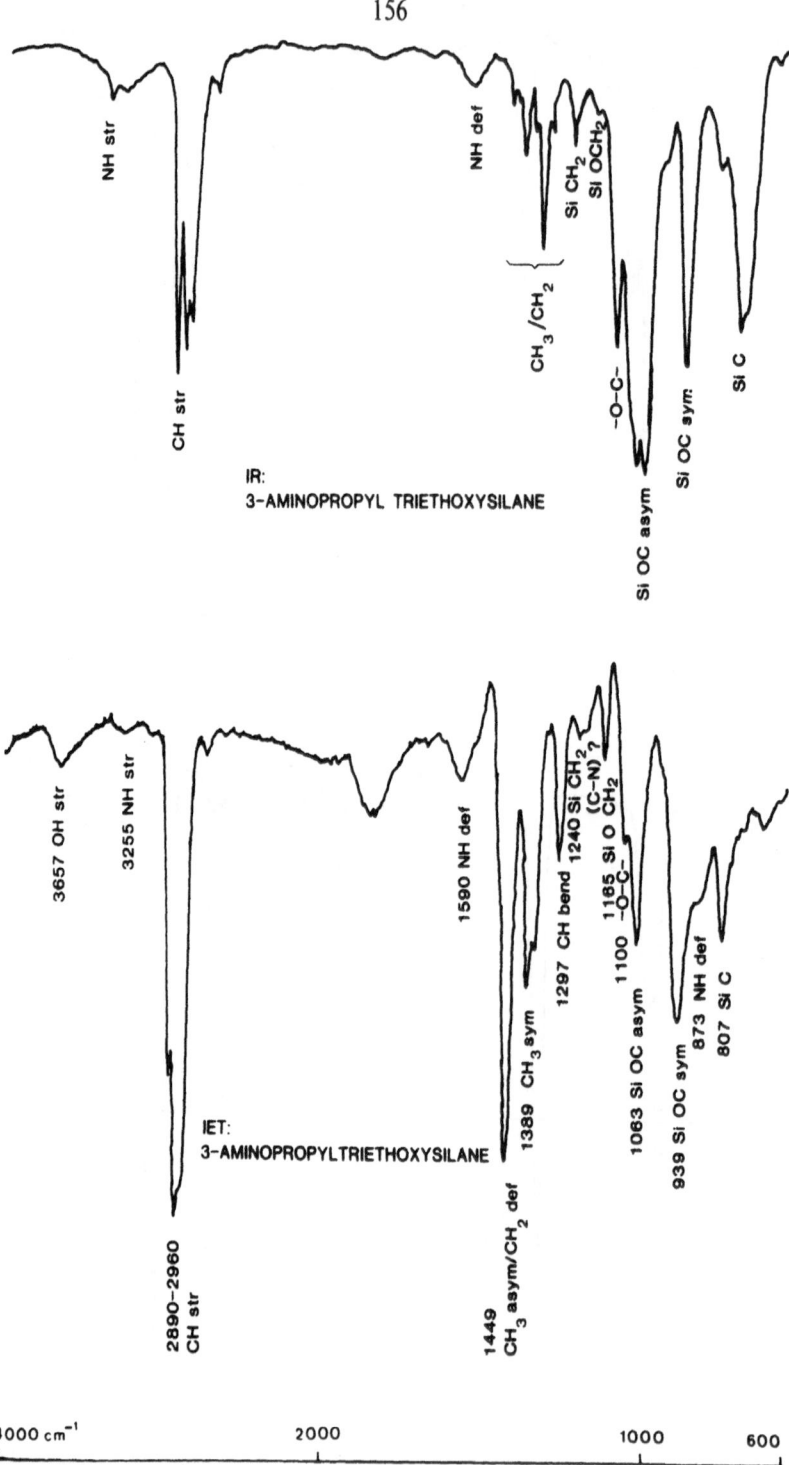

Figure 9 — IR and IET Spectra of 3-aminopropyl triethoxysilane.

peak at 1240 cm^{-1} could possibly be made up from the Si-CH and C-N vibrational modes and the shoulder at 873 cm^{-1} is the NH deformation mode.

It appears that when the triethoxysilanes are vapour phase doped onto aluminium oxide surface they are not chemically modified to any great extent. As the Si-H and Si-C absorptions are strong it may be deduced that these groups are perpendicular to the oxide surface so implying that the molecules are attached to the surface by the three oxygens. In the case of 3-aminopropyl-triethoxysilane the NH stretch at 3255 cm^{-1} in the IET spectrum is at a lower frequency by about 50 cm^{-1} than in the i.r., which might indicate hydrogen bonding. This might arise from the interaction between the NH$_2$ group and the surface. However, it is also possible that this may be due to internal hydrogen bonding in the following manner.

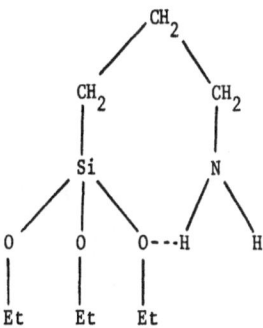

4.2 Trimethoxysilane

Ethyltrimethoxysilane (Fig. 10) shows little or no Si-O stretching and deformation modes, in IET spectra but the Si(C$_2$H$_5$) grouping gives very intense peaks at 952 cm^{-1}, 1008 cm^{-1} and 1237 cm^{-1} attributable to CH$_2$ and Si-CH$_2$ wagging modes. There are also two strong peaks at 1376 cm^{-1} and 1441 cm^{-1} assigned to CH$_3$ or CH$_2$ groups. Very weak peaks at 1066 cm^{-1} and 825 cm^{-1} are possibly due to Si-O-C asymmetric and symmetric modes. The peak of medium intensity at 690 cm^{-1} is due to CH$_2$ vibrations.

There are differences in the vinyl group modes in the IET spectra of vinyltrimethoxysilane and vinyltriethoxysilane. With the trimethoxy compound the intensity of the >C=C< and =CH$_2$ peaks at 1596 cm^{-1} and 3044 cm^{-1} suggests a higher degree of orientation for the vinyl group. The Si-O-C modes are very weak.

The IET spectrum of 3-glycidoxypropyltrimethoxysilane (Fig. 11) also shows major departures from the i.r. spectrum. These seem to include the presence of a >C=C< group at 1595 cm^{-1} and perhaps a carbonyl group at 1646 cm^{-1}.

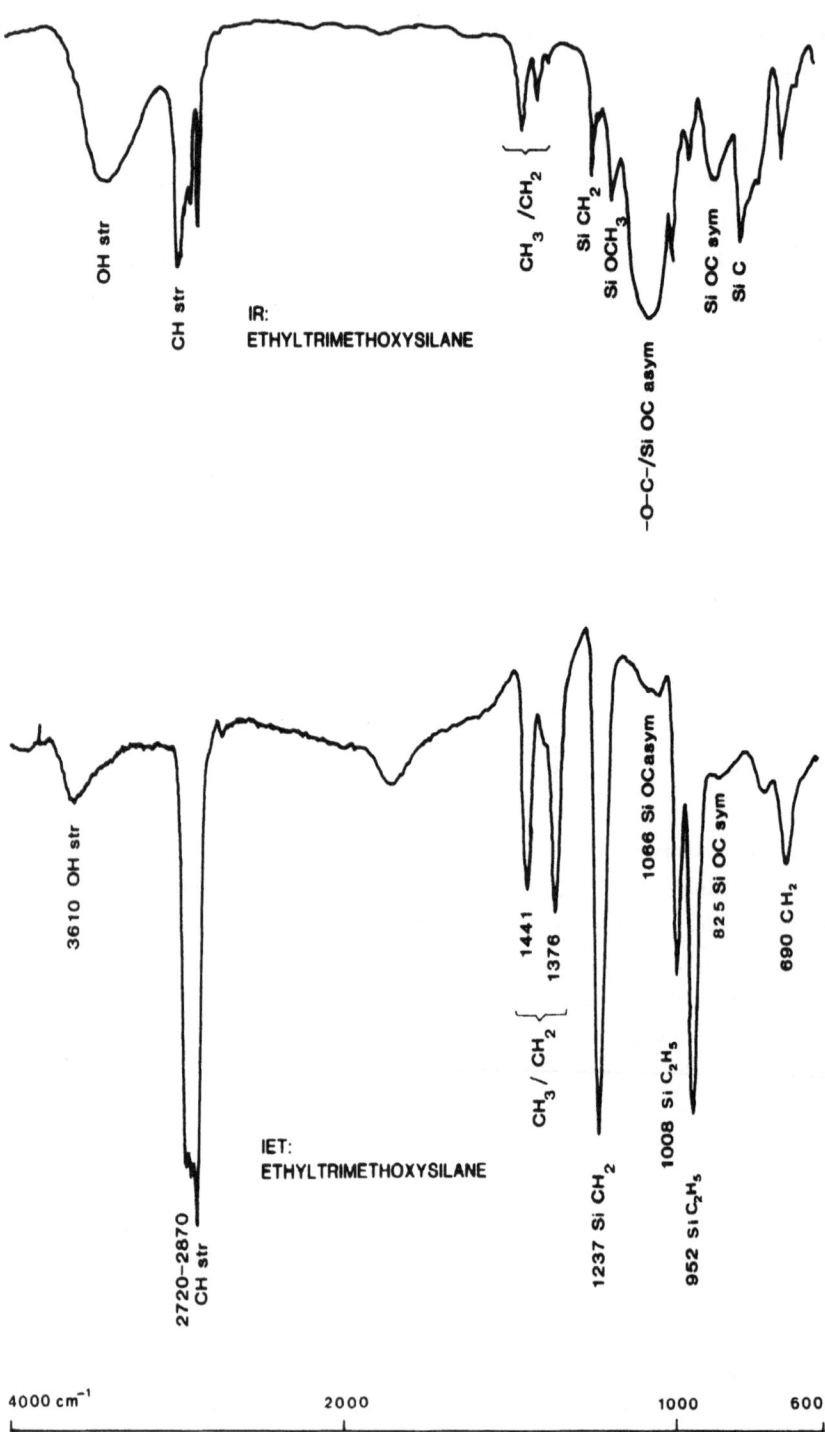

Figure 10 — IR and IET Spectra of Ethyltrimethoxysilane.

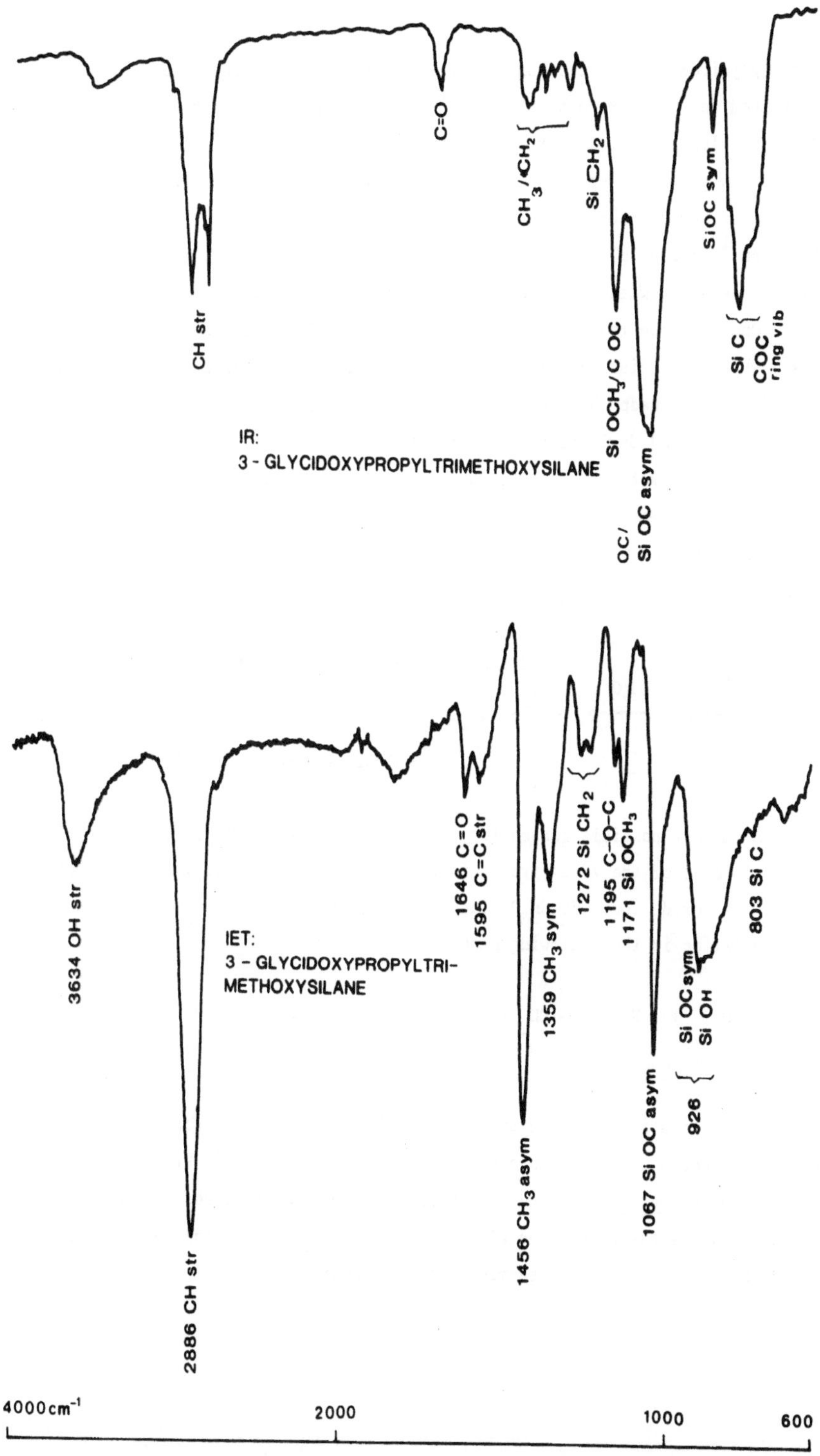

IR:
3 - GLYCIDOXYPROPYLTRIMETHOXYSILANE

CH str

C=O

CH₃/CH₂

Si CH₂

Si OCH₃/C OC

OC/
Si OC asym

SiOC sym

Si C
COC
ring vib

IET:
3 - GLYCIDOXYPROPYLTRI-
METHOXYSILANE

3634 OH str

2886 CH str

1646 C=O
1595 C=C str

1359 CH₃ sym

1456 CH₃ asym

1272 Si CH₂
1195 C-O-C
1171 Si OCH₃

1067 Si OC asym

Si OC sym
Si OH

926

803 Si C

4000cm⁻¹   2000   1000   600

Figure 11   -   IR and IET Spectra of 3-glycidoxypropyltrimethoxysilane.

Aged 3-glycidoxypropyltrimethoxysilane shows and i.r. absorption at 1730 cm^{-1} attributed to an ester carbonyl group[20]. The presence of substantial C-O-C groups is also suggested by the peak at 1195 cm^{-1}. Nevertheless it seems that the epoxide group in 3-glycidoxypropyl trimethoxysilane reacts within the junction or during fabrication, the evidence for this being the absence of peak at 939 cm^{-1} in the IET spectrum (Fig. 11). The situation with the epoxide groups contrasts sharply with its behaviour in epoxide-amine mixtures discussed earlier.

One possible reason for this is that chemical reactions occur with this silane which give rise to >C=O and >C=C< groups. A >C=C< stretching vibration at 1650 cm^{-1} has been observed by Allen, Hansrani and Wake[21] in the i.r. spectrum of acid hydrolysed 3-glycidoxypropyltrimethoxysilane and it was proposed that this arose by hydrolysis of the epoxide to a 1,2-diol followed by dehydration to a vinyl group. However, such a course would lead to an enol which would rearrange to the keto form and hence be a route to the observed carbonyl absorption.

A further possible reason for the disappearance of the epoxide groups and the development of the >C=C< group, is the anionic polymerisation of the epoxide groups. Hydroxide ions, or CH_3O^- ions produced by hydrolysis of silanes could cause initation in the following manner.

Polymerisation would occur by the addition of further epoxide rings and termination could give rise to carbon-carbon double bonds. St Pierre and Price[22] and Dege et al.[23] found i.r. absorptions due to allyl and propenyl ethers in polypropylene oxide prepared with alkaline catalysts and proposed that they were formed by the following termination mechanism.

$\wedge\!\wedge OCH_2 - CH \quad CH_2 \quad \rightarrow \quad$ or $\quad + \quad HOCHCH_2O^-$

(reaction scheme showing epoxide cleavage products: $\wedge\!\wedge OCH_2 - CH = CH_2$, $\wedge\!\wedge OCH = CH - CH_3$, and $HOCHCH_2O^-$ with CH_3 groups)

5 CONCLUSIONS

IET spectroscopy can be used to study how adhesive and adhesion promoters interact with metal oxides, and the technique has revealed interesting features in the two areas presented here. These features are the persistance of the epoxide group in epoxide adhesives, which is in contrast to its complete absence in the spectrum of the silane coupling agent containing epoxide groups. Other features of silane spectra are that chemical changes do not seem to occur when triethoxysilanes are absorbed on aluminium oxide in IET junctions whilst the trimethoxysilanes appear to react.

REFERENCES

1   Jaklevic, R.C. and Lambe, J., Phys. Rev. Lett., . 17, 1139, (1966).

2   Hansma, P., Phys. Rep., 30C, 145, (1977).

3   Hansma, P.K., "Tunnelling Spectroscopy", Plenum Press (1982).

4   Oxley, D.P., Bowles, A.J., Horley, C.C., Langley, A.J., Pritchard, R.G. and Runnicliffe, D.L., Surf. Inter. Anal., 2, 31, (1980).

5   Comyn, J., Horley, C.C., Oxley, D.P., Pritchard, R.G. and Tegg, J.L., J. Adhesion, 12, 171, (1981).

6   Brewis, D.M., Comyn, J. and Fowler, J.R., Polymer, 20, 1548, (1979).

7   Boerio, F.J., Polymer Prepr. (Amer. Chem. Soc. Div. Polym. Chem.), 22, 297 (1981).

8   Boerio, F.J., Cheng, S.Y., Armogan, L., Williams, J.W. and Gosselin, G., Proc. Annu. Conf. - Reinf. Plast./Compos. Inst; Soc. Plast. Eng. 35th, 23C (1980).

9   Boerio, F.J. and Williams, J.W., Appl. Surf. Sci., 7, 19 (1981).

10 Boerio, F.J. and Williams, J.W. and Burkstrand, J.M., J. Colloid Interface Sci., 91, 485 (1973).

11 Boerio, F.J., Polymer Prepr. (Amer. Chem. Soc. Div. Polym. Chem.) 24, 204 (1983).

12 Sung, N.H., Ni, S. and Sung, C.S.P., Org. Coat. Plast. Chem., 42, 743 (1980).

13 Sung, N.H. and Sung, C.S.P., Proc. Annu. Conf. - Reinf. Plast./Compos. Inst; Soc. Plast. Eng. 35th, 23B (1980).

14 Sung, N.H. and Kaul, A., Chin, I. and Sung, C.S.P., Polymer Eng. Sci., 22, 637 (1982).

15 Gettings, M. and Kinloch, A.J., J. Mater. Sci., 12, 2511 (1977).

16 Chiang, C.H., Liu, N-I. and Koenig, J.L., J. Colloid Interface Sci., 86, 26 (1983).

17 Shih, P.T.K. and Koenig, J.L., Mater. Sci. Eng., 20, 145 (1975).

18 Bascom, W.D., Macromolecules, 5, 792 (1972).

19 Brewis, D.M., Comyn, J., Oxley, D.P., Pritchard, R.G., Reynolds, S., Werrett, C.R. and Kinloch, A.J., Surf. Interface Anal., 6, 40,(1984).

20 Barratt, A.J., unpublished work (1980).

21 Allen, K.W., Hansrani, A.K. and Wake, W.C., J. Adhesion, 12, 199 (1981).

22 St. Pierre, L.E. and Price, C.C., J. Amer. Chem. Soc., 78, 3432 (1956).

23 Dege, C.J., Harris, R.L. and Mackenzie, J.S., J. Amer. Chem. Soc., 81, 3374 (1959).

Chapter 11

MONITORING ADHESIVE SOLVENT VAPOURS IN THE FOOTWEAR INDUSTRY

M F Denton

SATRA Footwear Technology Centre, Rockingham Road, Kettering, Northants

1 INTRODUCTION

The Footwear Industry relies heavily on the use of solvent based adhesives. It is, therefore, important to monitor solvent vapour levels to ensure that toxic hazards are minimised. Attention has always been given to the extraction of Solvent Vapours from the work areas around operations involving the use of adhesives and other solvent based preparations. However, only limited monitoring of vapour levels has been possible as only simple monitoring devices (eg direct reading detector tubes) were available. Recently more sophisticated devices have been developed and it therefore seemed appropriate to carry out a comprehensive survey of solvent vapour levels in shoe factories. At the same time a check was carried out on the efficiency of extraction devices, where provided, by measuring the air flow into them.

2 USES OF SOLVENT BASED PREPARATIONS IN THE FOOTWEAR INDUSTRY

Tables 1 and 2 list the various uses of adhesives and other solvent based preparations in the Footwear Industry and the types of solvent vapours that are likely to be found.

Table 1 :

Footwear manufacture - Areas using solvent based adhesives

Area of operation	Possible solvents present
Sole cementing	Acetone, MEK, Ethyl acetate, Toluene, Hexane,
Sole laying (welted)	SBP's, Dichloromethane, 1,1,1-Trichloroethane
Upper(shoe bottom) cementing	Acetone, MEK, Ethyl Acetate, Toluene, Hexane
Upper fitting	Acetone, MEK, Ethyl acetate, Toluene, Hexane, SBP's, 1,1,1-Trichloroethane
Solvent cement lasting	Acetone, MEK, Ethyl acetate, Toluene, Hexane
Heel covering (adhesive dip)	Acetone, MEK, Ethyl acetate, Toluene, Hexane
Veneers rand attachment and sole unit building	Acetone, MEK, Ethyl acetate, Toluene, Hexane
Lining lamination	Acetone, MEK, Ethyl acetate, Toluene, Hexane, SBP's, 1,1,1-Trichloroethane
Shank attachment	Acetone, MEK, Ethyl acetate, Toluene, Hexane

Table 2 :

Footwear manufacture - Areas involving solvent usage (other than adhesives)

Area of operation	Type of preparation	Possible solvents present
Sole priming	Wipes or primers	Acetone, MEK, THF, Ethyl acetate, Dichloromethane,Isopropanol,Ligroin,SBP's
Upper (shoe bottom) preparation	Wipes or primers	Acetone, MEK, THF, Ethyl acetate
Sole finishing	Finishes	Acetone, MEK, Ethyl Acetate, Butyl acetate, Cellosolve, Cellosolve acetate
Upper cleaning	Cleaners	Ligroin, SBP's, Methylated spirits, Isopropanol
Upper spraying	Lacquers	Acetone, MEK, Ethyl acetate, MIBK, Butyl acetate, Cellosolve, Cellosolve acetate
Heel cleaning	Wipes or dips	Ligroin, SBP's
Heel lacquering	Lacquers	Acetone, MEK, Ethyl acetate, Butyl acetate, Cellosolve, Cellosolve acetate, Toluene
Toe puff & stiffener	Activators	Acetone, MEK, Toluene, Dichloromethane
Bottom filling	Bottom fillers	Acetone, MEK

3 DEFINITIONS OF THRESHOLD LIMIT VALUES

The best criteria for acceptable solvent vapour levels in factory atmospheres are the threshold limit values (TLVs).Threshold Limit Values[1] refer to airborne concentrations of substances and represent conditions under which it is believed that nearly all workers may be repeatedly exposed day after day without adverse effect.

Two categories of Threshold Limit Values (TLVs) have been considered :
a. Threshold Limit Value-Time Weighted Average (TLV-TWA) : the time-weighted average concentration for a normal 8-hour workday or 40-hour workweek, to which nearly all workers may be repeatedly exposed, day after day, without adverse effect.

b. Threshold Limit Value-Short Term Exposure Limit (TLV-STEL) : the maximal concentration to which workers can be exposed for a period up to 15 minutes continuously without suffering (1) irritation, (2) chronic or irreversible tissue change, or (3) narcosis of sufficient degree to increase accident proneness, impair self-rescue, or materially reduce work efficiency, provided that no more than four excursions per day are permitted, with at least 60 minutes between exposure periods, and provided that the daily TLV-TWA also is not exceeded. The STEL should be considered a maximal allowable concentration, or ceiling, not to be exceeded at any time during the 15 minute excursion period.

The definitions of time-weighted average allow the concentrations to rise above the limit for short periods provided they are compensated by equivalent periods below the limit during the workday. In some instances it may be permissible to calculate the average concentration for a workweek rather than for a workday.

In spite of the fact that serious injury is not believed likely as a result of exposure to the threshold limit concentrations, the best practice is to maintain concentrations of all atmospheric contaminants as low as is practical.

3.1 Mixtures

When two or more hazardous substances are present, their combined effect, rather than that of either individually, should be given primary

consideration. In the absence of information to the contrary, the effects of
the different hazards should be considered as additive. That is, if the sum
of the following fractions,

$$\frac{C_1}{T_1} \quad \frac{C_2}{T_2} + \ldots \frac{C_n}{T_n}$$

exceeds unity, then the threshold limit of the mixture should be considered as
being exceeded. C_1 indicates the observed atmospheric concentration, and T_1
the corresponding threshold limit.

Example
Air contains 400ppm of acetone (TLV = 1000ppm), 150ppm of sec-butyl acetate
(TLV = 200ppm) and 100ppm of 2-butanone (TLV = 200ppm).

Atmospheric concentration of mixture = 400 + 150 + 100 = 650ppm of mixture

$$\frac{400}{1000} + \frac{150}{200} + \frac{100}{200} = 0.4 + 0.75 + 0.5 = 1.65$$

Threshold limit is exceeded.

4 REVIEW OF MONITORING DEVICES
        Before carrying out the survey it was necessary to review the various
types of monitoring device available and choose which to use. It was found
that monitoring devices could be broken down into two broad groups. These are
direct monitoring devices, which provide on-site (spot) measurements of
solvent vapour levels, and sampling devices, which absorb a representative
portion of the vapours in the atmosphere for later analysis in the
laboratory.

4.1   Direct Monitoring Devices
4.1.1 Direct reading detector tube : This method uses tubes containing
reagents which give a colour change in the presence of specific contaminants.
A hand-operated pump (fig. 1) draws a known volume of air through the tube
which is calibrated so that the length of colour change can be related to the
level of contamination. Tubes are available for a wide range of solvent
vapours, but interference from similar vapours can be a problem.

Although the method is designed chiefly for spot measurements of individual
vapour components, tubes do exist that can be used for average concentrations

over a period for some vapours. These require to be used with a personal sampling pump (Section 4.2.1).

The use of these tubes is really only helpful where nearly all the risk arises from a single known vapour and in the absence of interfering vapours of differing risk. This situation rarely arises in shoe factories.

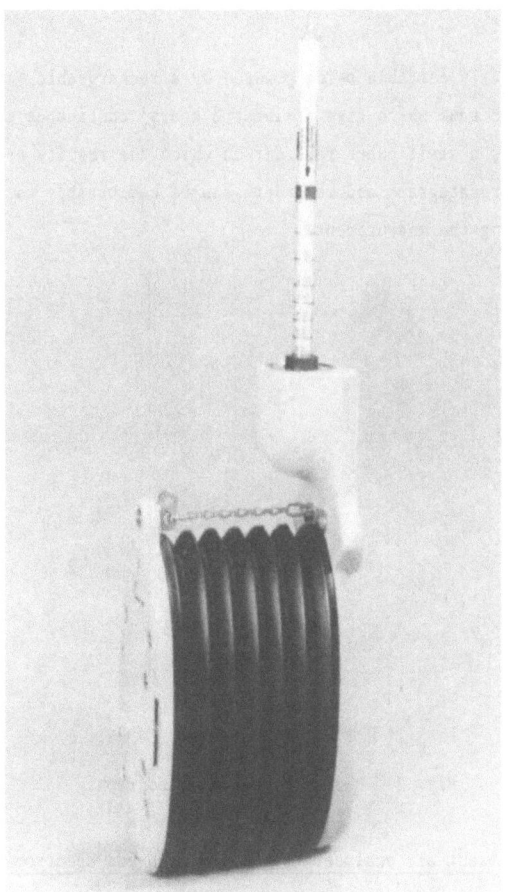

Fig. 1   Hand operated pump and direct reading detector tube

4.1.2. Organic vapour analyser (portable gas chromatograph) : The instrument used (fig. 2) is fitted with a flame ionisation detector. Air is drawn into the instrument by means of a small electrically driven pump. In one mode of operation the sample is passed directly into the detector thus giving a continuous readout of total organic vapour present. The reading is given in 'methane equivalent', correction factors specific for the components present (typically about 1.25) must be applied. Alternatively an increment of the sample can be passed through a chromatographic column prior to entering the detector; this gives a measure of the individual sample components present at the time the sample was taken.

The unit is completely portable being powered by a rechargeable battery and refillable hydrogen tank which give at least 8 hours' continuous use. The system incorporates a small chart recorder on which the results are plotted. The whole unit, chromatograph and recorder, can be comfortably worn by straps by the person making the measurement.

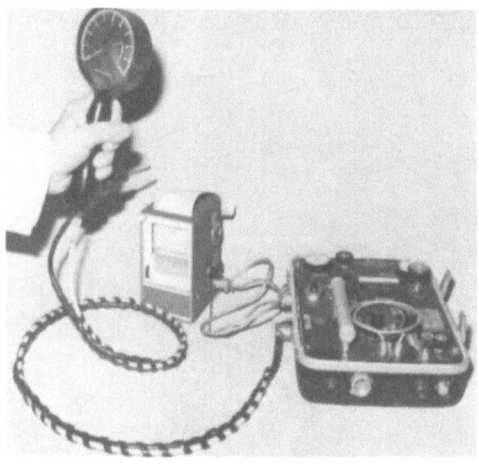

Fig. 2   Portable gas chromatograph

4.1.3 <u>Portable ambient air analyser (portable infra-red spectrometer)</u> : This instrument gives on-site measurements of vapour concentration by passing vapours through a gas cell and obtaining an infra-red spectrum. Some substances have a characteristic colour due to absorption in the visible light spectrum, and most substances absorb at characteristic wavelengths in the infra-red; thus measurements of the absorption at these wavelengths can be related to concentration. Continuous monitoring is possible if a chart recorder is also used.

4.1.4 <u>Portable gas analyser (photo-ioniser)</u> : On-site measurements are
obtained by passing air into a chamber where it is exposed to UV light and the
conductivity of the air is monitored. The presence of organic vapours causes
a change in the conductivity of the air by creating ions; and this change is
related to the concentration of organic vapour by the use of calibration
graphs. Again, continuous monitoring is possible if a chart recorder is used.

4.2   Sampling Devices

4.2.1 <u>Personal sampling pump</u> : A small battery-operated pump, light enough to
be worn by an operative, is used to draw air into a tube through an adsorbing
material, such as charcoal, and the tube is then returned to the laboratory
for analysis by gas liquid chromatography. If sampling time and pump-flow
rate are known, the average concentration can be calculated. The method
allows monitoring of individual vapours; but sampling for a full 8-hour shift
would require a large amount of adsorbent.

Either activated charcoal or polymeric adsorbents may be used in the tube.
Desorption can either be achieved thermally or by the use of an appropriate
solvent. In this project charcoal packings and desorption by carbon
disulphide were used exclusively.

This is the technique currently recommended by the American National Institute
for Occupational Safety and Health (NIOSH) even though short sampling periods
are necessary - typically less than one hour.

4.2.2 <u>Organic vapour dosimeter (badge)</u> : This is a lightweight plastic badge
(Fig. 3) which is worn by the operative. The badge samples the atmosphere
throughout a working shift, by diffusion across a measured air gap onto a
charcoal collection element.

At the end of the shift the badge is sealed in a foil packet which is marked
with the time worn. The badge is returned to the laboratory where the
contaminants are desorbed from the collection element with carbon disulphide
and analysed by gas liquid chromatography. The desorption is carried out in
the badge itself, the diffusive cover being replaced by a plastic analysis
cover with plugged vents for the addition of carbon disulphide and removal of
samples for GLC. The time weighted average vapour concentrations for a full
working shift are thus obtained.

Fig. 3   Organic vapour dosimeter

Such badges are said to be valid only on the presence of a minimum air movement, which is adequately met by wearing of the badge.

4.2.3 <u>Tube type diffusive sampler</u> : A tube about the size of a fountain pen and containing a polymeric adsorbing element is worn by an operative, and used in a similar manner to the dosimeter. Although in this case the vapours trapped on the element are desorbed by the use of heat. The analysis is carried out using an automatic sample handling device linked to the gas liquid chromatograph.

4.2.4 <u>Colorimetric method</u> : A known volume of an atmsophere is drawn through an absorbing solution which is then treated with a material which reacts with a particular contaminant.   The amount of colouration which results can be measured by a colorimeter and related to the amount of contaminant vapour in the atmosphere. (This method has been mainly used for spot measurements of individual vapours, but could probably be adapted for period averages).

4.3   <u>Chosen Techniques</u>
       The techniques chosen for use in the factory survey are listed below :

1. Direct reading detector tube

2. Organic vapour analyser (portable gas chromatograph)

3. Personal sampling pump + charcoal filled tubes

4. Organic vapour dosimeters (badges)

5   VERIFICATION OF TECHNIQUES

Before commencing actual measurements of solvent vapour levels in shoe factories it was necessary to verify that the various techniques chosen gave meaningful results. In order to do this it was necessary to construct an apparatus in which atmospheres containing concentrations of solvent vapours could be generated.

After consultation with the Health and Safety Executive (Cricklewood Laboratories) an atmosphere generator (shown in Fig. 4) was constructed.

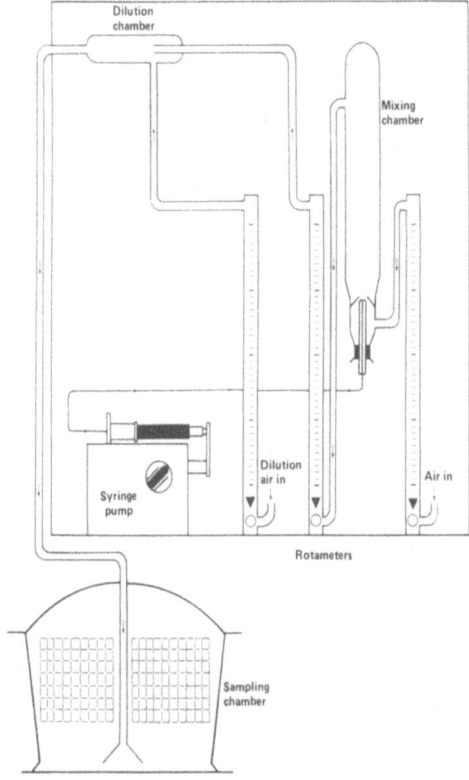

Fig. 4   Standard atmosphere generator

This consists of a motor driven gas tight syringe which is used to inject the liquid solvent at a known constant rate into a mixing chamber where it is atomised in a regulated supply of air simultaneously applied to the atomisation chamber. If desired the atmosphere generated in the mixing chamber is diluted with a second regulated supply of air. The concentration of atmosphere generated is calculated from the rate of injection of the solvent and the rate of supply of air.

This device was used to assess the performance of the various sampling devices. Calibration factors were established for the organic Vapour Analyser. The badge type organic Vapour Dosimeters were shown to be accurate to $\pm$ 10%.

The verification of all other solvent vapour measuring techniques used is well documented. These include Personal Sampling pumps plus Niosh tube[2] and Direct Reading Detector tube[3].

6   RESULTS OF SURVEY

Forty-two factories were tested during the survey. The average number of positions tested being approximately six per factory. Seventeen different operations were covered. Table 3 shows the results obtained. Table 4 breaks these results down by type of operation.

Table 3 :
Overall results

	Monitored	Over limit *
Factories	42	20 **
Positions	244	39 *

*   Based on best available results (either Organic Vapour Dosimeter or
    Organic Vapour Analyser)
** In only isolated positons

Table 4 :

Results - Breakdown by operation

Operation	Monitored	Over limit
Bottom filling	3	0
Bottom cementing	42	11
Closing (fitting)	33	4
Sole priming	12	4
Sole cementing	32	7
Sole prime & cement (combined)	8	3
Sole coating (for sole laying - welted)	2	0
Insole cleaning	3	0
Insole cementing	7	1
Heel dipping	4	2
Heel covering	5	1
Coating heel veneers	3	0
Edge finishing	4	0
Finish spraying	21	0
Shoe rooming	8	1
Upper cleaning	9	1

7   VENTILATION SURVEY RESULTS

The study of ventilation in shoe factories was limited to measurement of air flow into various extraction devices. This was done by using a portable hot wire anemometer. Measurements were of extraction velocity which should correspond to capture velocity when an extraction device, such as a down draught grill, is being properly used. Up to the present some recommendations of capture velocity in the shoe industry has been 16m/min (0.27m/s) but the Health and Safety Executive have suggested a value of 24m/min (0.4m/s).

Table 5 shows the results of the ventilation survey, broken down according to type of extraction device. The results are then compared to recommended capture velocities.

The practical situation appears to be that when reasonably designed extraction equipment exists it usually meets the value of 24m/min (0.4m/s).

Table 5 :
Results of ventilation survey

Type of extract-ion device	Number monitored	Range of air flow (m/s)	Number below 0.27m/s	No. between 0.27m/s and 0.40m/s	Number above 0.4m/s
Down draught extraction bench	17	0.02–1.44	3	1	13
Spray booth	20	0.25–2.83	2	1	17
Drying cabinet	6	0.05–1.49	5	0	1
Extraction device fitted to machine	6	1.00–6.00	0	0	6

8   CODE OF PRACTICE

As a result of the survey a Code of Practice for Safe Handling of Solvents and Adhesives in the Footwear Industry was drawn up. This recommends that checks of solvent vapour levels are carried out at six monthly intervals.

9   MONITORING SERVICES

Two routine solvent vapour monitoring services for shoe manufacturers have also been set up. These utilise the portable gas chromatograph and the badge type organic vapour monitors. The results to date of the routine badge monitoring service (160 positions monitored) show that at approximately 1 in 6 of these positions the solvent vapour levels exceed the acceptable limit (TLV-TWA).

REFERENCES

1. Health and Safety Executive Guidance Note EH15. Threshold Limit Values

2. Niosh. D G Taylor, R E Kupel, J M Bryant - Documentation of the Niosh
    Validation Tests, US Department of Health, Education and Welfare 1977

3. Draeger. Title 42 - Public Health Service, Department of Health,
    Education and Welfare . Part 84 - Certifcation of Gas Detector Tube
    Units. Federal Register, 38 (1973) No. 88

Chapter 12

SPONTANEOUS SHEAR DELAMINATION OF ADHESIVE JOINTS

M. BARQUINS

E.R. Mécanique des Surfaces du C.N.R.S.
Laboratoire Central des Ponts et Chaussées, Paris, France

1 INTRODUCTION

Most thin films, as paints, varnishes and coatings, have the drawback of flaking and peeling off when the substrate is deformed. A crack, initiated from a surface flaw, propagates at the interface and causes a partial rupture of facing. In order to better understand this behaviour, Kendall[1], in 1975, has investigated the problem of the failure of lap shear joints using an energy balance concept theory based on optimizing the total energy of the system at the equilibrium. The equilibrium state is reached if the first derivative of the total energy, with respect to the contact area, is cancelled, and the corresponding force is called elastic adherence force. So, one can deduce that a lap shear joint cannot sustain a prestress greater than its normal adherence force.

The energy balance concept theory has been successfully applied by Kendall to determine the crack initiation force in peeling (ref.2), the critical force of interfacial failure in laminates (ref.3) and composites (ref.4) and the adherence force of rigid spherical (ref.5) and flat (ref.6) punches in adhesive contact with elastomers. But, as pointed out by Maugis and Barquins[7] in 1978, this theory cannot give information about the stability of the system that depends on the second derivative of the total energy. In these conditions, Maugis and Barquins[7] were led to reintroduce the concepts of fracture mechanics, such as the strain energy release rate G, and to study the stability according to the sign of the derivative of G with regard to the contact area. This approach has the advantage of enabling one to study the kinetics of crack propagation and to predict the evolution of the system, whatever the geometry of contact and experimental procedure.

The general equation of the kinetics of the adherence of elastomers, proposed by Maugis and Barquins[7]:

$$G - w = w.\Phi(a_\theta.v) \qquad \qquad \cdots (1)$$

where w is the thermodynamic work of adhesion and $\Phi(a_\theta.v)$ a dissipation function characteristic of the viscoelastic material and of the propagation in mode I, has been verified for kinetics of adherence of rigid spherical and flat punches in adhesive contact with polyurethane plates. All the theoretical predictions were confirmed in four different push on/pull off tests: fixed load (ref.7), fixed cy-clic loading (ref.8), fixed displacement (ref.9) and fixed crosshead velocity (ref.10), this last point solving the problem of the tackiness of elastomers.

The purpose of the work reported here was to show that eq. 1, experienced in normal approach measurements of the adherence, remains valid in shear delami-nation geometry and allows one to wholly describe the spontaneous peeling of rub-ber-like materials in adhesive contact with a deformable substrate. Particularly, it is shown that eq. 1 enables one to predict the time of complete rupture of the lap joint and the values of the corresponding force brought into play, when the substrate is submitted to instantaneous or increasing tensile elongations.

2 ADHERENCE OF VISCOELASTIC SOLIDS

Lately, Maugis and Barquins[11] have shown that the contact of two elastic bodies can be treated as a thermodynamic problem. For instance, let us consider two elastic solids in adhesive contact over an area A under a compressive or ten-sile load P that causes an elastic displacement δ perpendicular to the interface (fig.1). A variation of the contact area can be realized in a test at fixed load or at fixed grips conditions, so that the state of the system depends upon two independent variables P and A or δ and A. Regarding the edge of the contact area as a crack tip propagating in mode I, that recedes or advances accordingly as the area of contact decreases or increases, Maugis and Barquins[11] were led to reintro-duce the concepts of fracture mechanics, such as the strain energy release rate G. G is calculated from the elastic stored energy U_E and the potential energy U_P of the load by $G=(\partial U_E/\partial A)_P+(\partial U_P/\partial A)_P$ in a test at fixed load, and by $G=(\partial U_E/\partial A)_\delta$ in a test at fixed grips conditions. Neglecting thermal effects, variations of the Helmotz and Gibbs free energies of the system for a fluctuation dA of the contact area are respectively: $dF=Pd\delta+(G-w)dA$ and $dG=-\delta dP+(G-w)dA$, where w is the thermo-dynamic work of adhesion defined from the surface and interface energies of so-lids 1 and 2 by $w=\gamma_1+\gamma_2-\gamma_{12}$. The equilibrium state of the system corresponds to the maximum of F or G, following the experimental procedure ($d\delta=0$ or $dP=0$), so that it is defined by $G=w$ (Griffith criterion). So, the relations giving the strain energy release rate and the elastic displacement as a function of the contact area and the load appear as the equations of state of the system. The equilibrium is stable if $\partial G/\partial A$ is positive, unstable if negative, the elastic adherence force being the force corresponding to the borderline $\partial G/\partial A=0$.

This equilibrium may be disrupted by a change in load or in elastic displacement, and if $G \neq w$, the area of contact spontaneously varies in order to decrease thermodynamic potentials. If $G < w$, the contact area increases and the crack recedes and conversely, if $G > w$, the contact area decreases and the crack advances. When the crack extends by dA, the mechanical energy released is GdA, whereas wdA is the amount of energy required to break interfacial bonds, so that the excess $(G-w)dA$ is turned into kinetic energy, if there is no dissipative factor. For viscoelastic solids, as rubber-like materials, the difference $G-w$ is the crack extension force applied to the crack tip; under this force, the crack takes a limiting speed v that depends on the temperature. If it is accepted that viscoelastic losses are proportional to the thermodynamic work of adhesion w, as shown by Gent and Shultz[12] from peeling experiments on liquids, and by Andrews and Kinloch[13] from theoretical considerations and experiments with rigid substrates; and assuming that these viscoelastic losses are localised at the crack tip, one can write[7]:

$$G - w = w . \Phi(a_\theta . v) \qquad \qquad . . . (1)$$

where the second term is the viscous drag resulting from the losses limited to the crack tip. $\Phi(a_\theta . v)$ is a dimensionless function of the crack speed v and of the temperature through the shift factor a_θ of the Williams-Landel-Ferry[14] transformation. We have shown that Φ is independent of the loading system and of the geometry of contact. This function Φ is characteristic of the viscoelastic material for the propagation in mode I and may be directly linked to the frequency dependence of the imaginary component of the Young modulus. Knowledge of the function Φ makes it possible to predict the evolution of the contact area in all circumstances. The prediction assumes only that rupture is adhesive, i.e. that the crack propagates at the interface, and the application of eq. 1 implies that gross displacements are purely elastic, with G computed from the relaxed elastic modulus E_o and that the frequency dependence of E only appears at the crack tip, where deformation velocities are high. The main interest of eq. 1 is that surface properties (w) and viscoelastic losses (Φ) are clearly decoupled from elastic properties, geometry and loading conditions that only appear in G.

Numerous meticulous adherence experiments carried out with glass balls, flat glass punches and flat ended glass balls in adhesive contact with polyurethane plates in push on/pull off tests at fixed load (refs.7,15,16), cyclic loading (ref.8), fixed displacement (ref.9) and fixed crosshead velocity (ref.10), verify all theoretical predictions. Whatever the intrinsic properties (surface and viscoelastic) of tested solids or such experimental parameters (geometry of contact, speed of separation, temperature, relative humidity, initial applied load which presses together the two solids and duration of this initial contact), all experimental points fall on a master curve in a diagram $\Phi(v)$. It has been shown that over a large range of speeds of propagation (10^{-1}-10^{4} μm/s), the function Φ may be represented by the power function:

$$\Phi(a_\theta \cdot v) = \alpha(\theta) \cdot v^n \qquad \qquad \ldots (2)$$

where $\alpha(\theta)$ is a parameter depending on the temperature (ref.10) and n is a characteristic of the rubber-like material and takes the value 0.6 for polyurethane. One can point out that the model proposed by Greenwood and Johnson[17], for a three-element viscoelastic material, gives a variation of G with the 0.5th power of the crack speed which is in reasonably good agreement with our experimental value. Lastly, the multiplicative effect of w on viscoelastic losses, seen in eq. 1, that can only arise if the interface is capable of withstanding stresses (ref.13), has been confirmed by cylinder rolling experiments in an atmosphere of variable humidity (refs.18,19).

In the work reported here, we show that eq. 1, written and experienced for crack propagation in mode I, also allows one to study the spontaneous shear delamination of rubber strips in contact with deformable substrates.

THEORY OF THE SPONTANEOUS PEELING

Let us consider a lap shear joint (fig. 2), almost similar to that investigated by Kendall[20] in 1975, made by contacting two smooth strips of rubber-like materials. The upper strip (Young modulus E_2, thickness h_2, width b and length L) is in adhesive contact on to an area A=bL with the lower strip used as substrate (Young modulus E_1, thickness h_1, width b and length L+u). The two surfaces adhere under the only molecular attraction forces, without additional adhesive or pre-stress. Moreover, one extremity of the lap joint is clamped. In these conditions, if an instantaneous adequate tensile elongation is imposed to the substrate, a crack appears at the free end of the upper strip and propagates inwards along the interface (fig.2B), the spontaneous delamination being able to cause the complete rupture of the joint. At a given time t, the system being submitted to an elongation δ, let be x the length of the relaxed peeled part of the upper strip, and ℓ the length of the adhesive part of the joint (contact length). Assuming that the peeled part of the upper strip is unstrained and neglecting the distorsion of the stress field near the crack tip, the system can be schematized as a simple assembling of springs, two springs in parallel connection (adhesive part of the joint)

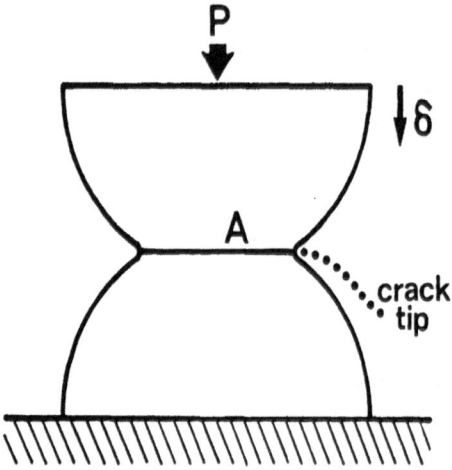

Fig. 1   Equilibrium contact of a rigid sphere on an elastic solid.

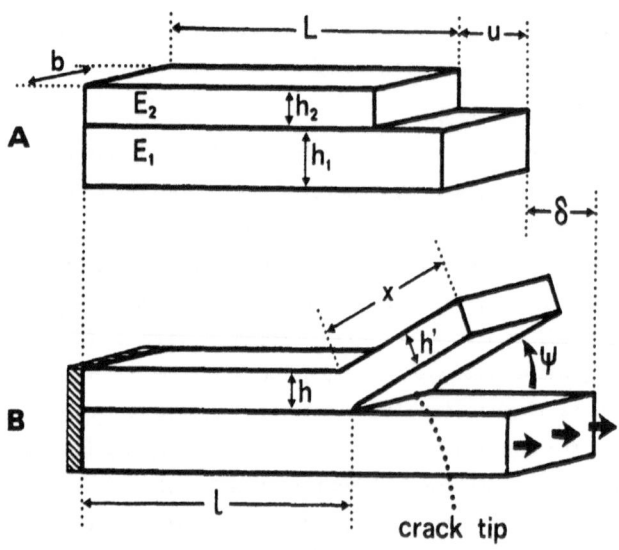

Fig. 2   Geometry of the lap shear joint tested. A :relaxed; B :strained.

in series with a third (free end of the substrate). So, an elementary application of Hooke's law allows one to calculate the force applied to the system:

$$F = \delta b E_1 h_1 (E_1 h_1 + E_2 h_2) / \left[E_1 h_1 (L+u) + E_2 h_2 (x+u) \right] \qquad \ldots (3)$$

From the stored elastic energy in the system, $U_E = F\delta/2$, one can easily deduce the strain energy release rate G by $G = (\partial U_E / \partial A)_\delta$:

$$G = \delta^2 E_1 h_1 E_2 h_2 (E_1 h_1 + E_2 h_2) / 2 \left[E_1 h_1 (L+u) + E_2 h_2 (x+u) \right]^2 \qquad \ldots (4)$$

So, the equilibrium relationship can be written:

$$\delta = \left[E_1 h_1 (L+u) + E_2 h_2 (x+u) \right] \cdot \left[2w / E_1 h_1 E_2 h_2 (E_1 h_1 + E_2 h_2) \right]^{\frac{1}{2}} \qquad \ldots (5)$$

The minimum value $\delta_{init.}$ of the elongation required to observe the crack initiation is given by eq. 5 with x=0. Substituting eq. 4 in eq. 3, the corresponding value of the force applied to the system is:

$$F_{init.} = b \left[2w E_1 h_1 (E_1 h_1 + E_2 h_2) / E_2 h_2 \right]^{\frac{1}{2}} \qquad \ldots (6)$$

a same formula has been already given by Kendall[20] and, as expected, this critical force is independent of initial lengths of substrate and upper strip.

The stability of the system is given by the sign of $(\partial G / \partial A)_\delta$; as this derivative is positive, the equilibrium is stable, i.e., if a fluctuation decreases the contact area A, the strain energy release rate G incrementally decreases and becomes inferior to the thermodynamic work of adhesion w, hence the crack recedes to its equilibrium position. The crack can only advance if the tensile elongation is varied bringing back G to the value w : one is dealing with the controlled rupture of an adhesive joint. This fact is clearly visible on fig. 3 A and B, which shows the strain energy release rate G as a function of the relaxed peeled part length x of the upper strip, in reduced coodinates, for different values of the instantaneous tensile uniaxial deformation imposed to the adhesive joint, and for the two symmetrical cases : $E_2 h_2 / E_1 h_1 = 5$ and $E_2 h_2 / E_1 h_1 = .2$. As soon as a fixed tensile elongation δ greater than $\delta_{init.}$, given by eq. 5 with x=0, is applied to the substrate, G takes the corresponding value given by eq. 3 and instantaneously a crack is initiated at the free end of the upper strip and propagates in the interface (fig. 2B). Then, x increases and G continuously decreases with time. The minimum value of the uniaxial strain to observe the initiation of the crack, for the practical case u=0, can be deduced from eq. 4 with x=0 :

$$\varepsilon_{init.} = \left[2w E_1 h_1 / E_2 h_2 (E_1 h_1 + E_2 h_2) \right]^{\frac{1}{2}} \qquad \ldots (7)$$

This relation proves that an adhesive joint, made by facing two dissimilar elastic strips, is stronger if the substrate is characterised by a parameter Eh greater than the peeled strip ones, as confirmed by fig. 3 A and B.

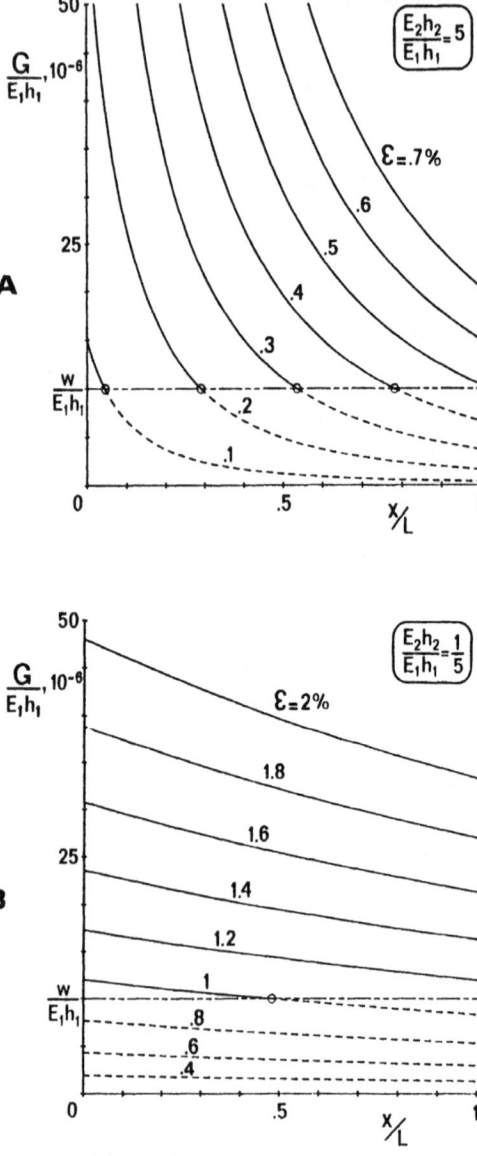

Fig. 3   Strain energy release rate versus length of the relaxed peeled strip, for various instantaneous tensile elongations, in reduced coordinates. A : $E_2h_2/E_1h_1 = 5$; B : $E_2h_2/E_1h_1 = .2$.

As soon as the crack is initiated ($\delta > \delta_{init.}$), fig. 3 shows that the propagation can lead to an equilibrium state if G takes the value w before the total rupture of the joint is reached, i.e. $x_{equil.} < L$, or to the complete loss of contact if G>w with x=L. Hence, it is possible to define a new critical value of the fixed elongation δ corresponding to this borderline case. For the practical situation u=0, one can deduce from eq. 5 with x=L, the minimum value of the instantaneous uniaxial deformation to cause an ineluctable complete rupture of the adhesive joint :

$$\varepsilon_{crit.} = \left[2w(E_1h_1+E_2h_2)/E_1h_1E_2h_2\right]^{\frac{1}{2}} \qquad \ldots(8)$$

Variations of the two particular deformations $\varepsilon_{init.}$ and $\varepsilon_{crit.}$ as a function of the E_2h_2/E_1h_1 ratio are shown on fig. 4. So, it is possible to predict, following the value of the ratio E_2h_2/E_1h_1 with w remaining constant, the conditions to observe no rupture and partial or complete rupture of an adhesive joint by spontaneous shear delamination if an instantaneous uniaxial deformation is applied to the substrate. Figure 4 clearly shows that a coating characterized by a low ratio E_2h_2/E_1h_1 can sustain a high instantaneous uniaxial strain. Obviously, if the ratio E_2h_2/E_1h_1 takes high values, the deformation $\varepsilon_{crit.}$ tends towards the well-known value of crack initiation in a standard peeling test at zero angle from a rigid substrate : $\varepsilon = \sqrt{2w/Eh}$, (ref.21).

From the simple application of the Hooke law to the adhesive part of the joint and to the strained free end of the substrate, one can write a relation including all the geometrical and mechanical parameters of the system, so that the length x of the relaxed peeled part of the upper strip may be determinated by the positive root of the equation :

$$x^2E_2h_2 + x\left[E_1h_1(L+u+\delta)-E_2h_2(L-u-\ell)\right] + E_1h_1\left[L(\ell-L-u-\delta)+\ell u\right]$$
$$+ E_2h_2u(\ell-L) = 0 \ldots(9)$$

of which the crack speed v=dx/dt, for the relaxed conditions, at fixed tensile elongation δ and at fixed crosshead velocity $\dot{\delta}$ can be deduced :

$$v_\delta = -\frac{d\ell}{dt}\left[E_1h_1(L+u)+E_2h_2(x+u)\right]/\left[E_1h_1(L+u+\delta)-E_2h_2(L-u-2x-\ell)\right] \quad \ldots(10)$$

$$v_{\dot{\delta}} = v_\delta + E_1h_1(L-x)\dot{\delta}/\left[E_1h_1(L+u+\delta)-E_2h_2(L-u-2x-\ell)\right] \qquad \ldots(11)$$

the derivative $d\ell/dt$ being calculated by the slope in every points of experimental curves of variations of the contact length ℓ with time.

So, the strain energy release G, given by eq. 4, with x calculated by eq. 9, can be related to the crack speed v assessed by eq. 10 or 11 following the experimental procedure, in order to verify the general equation (eq.1) previous-

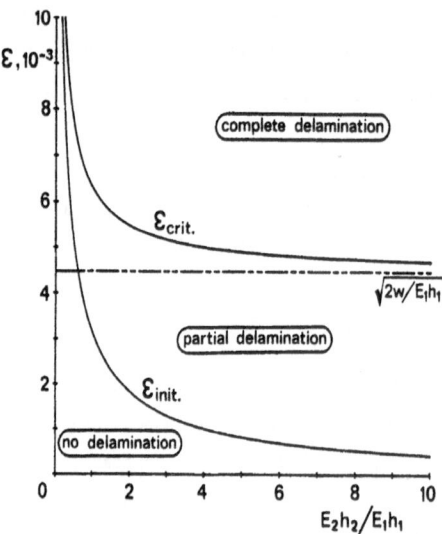

Fig. 4   Particular values of the instantaneous uniaxial deformation versus E_2h_2/E_1h_1 ratio, to observe initiation of the crack ($\varepsilon_{init.}$) and complete rupture of the contact ($\varepsilon_{crit.}$).

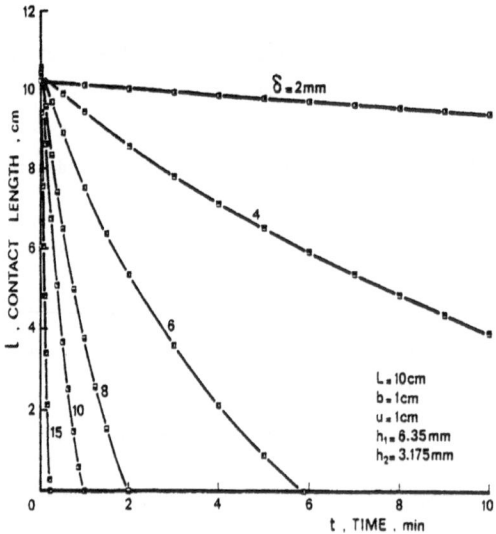

Fig. 5   Variations with time of the contact length between two strips of polyurethane, for various instantaneous tensile elongations applied to the substrate, as in fig. 2B.

ly proposed and experienced in normal approach measurements of the adherence of
rigid punches in adhesive contact with a viscoelastic material (refs. 7-10).

3 EXPERIMENTAL METHOD

Experiments were carried out using transparent optically smooth polyurethane (PSM4 Vishay, E=3.6 MPa), similar to that used for previous studies on the
kinetics of adherence of glass spherical or flat punches (ref.7), and unfilled
natural rubber (Impervia, E=1.3 MPa), already tested in friction experiments
(ref.22). They were delivered as plates of thicknesses h=3.175 mm and 6.35 mm
(PSM4) and 2 mm (NR). Before facing, the surfaces were wiped with an alcohol-
soaked rag, dried with warm air and left, sheltered from dust, during 30 min for
the equilibrium with room temperature to be reached. Then, two strips with various
lengths L, free end lengths u and widths b varying in the range 5-20 cm, 5-20 mm
and 0-10 cm respectively, were gently superimposed and they adhered under the only
molecular attraction forces, without additional adhesive or prestress. In order
to avoid dwell time effects (ref.16), strips were coupled during the same long
contact duration 30 min, for any set of experiments. Moreover, temperature (23°C)
and humidity (73%) were kept constant. It can be pointed out that the shear delamination producing no visible damage to the surfaces, just a small lot of samples
were used and have given reproducible values of crack speed better than 4%.

Experiments at fixed grips conditions and at fixed crosshead velocity
were carried out using a tensile machine (Instron 1026) that enabled us to impose,
with help of two gear boxes, speeds $\dot{\delta}$ varying in the range .5-500 mm/min, and to
measure forces to 50 N. For a quantitative evaluation of the crack speed during
the delamination, the crack tip being always visible through the transparent substrate, a 16 mm camera recorded the experimental arrangement at a rate varying
from 10 to 50 frames/s; the frames were then enlarged and the contact length ℓ
was measured, and the delamination angle ψ also (fig. 2B).

4 SPONTANEOUS DELAMINATION AT FIXED TENSILE ELONGATION

In a first set of experiments, we have tested the model proposed, studying
the kinetics of spontaneous peeling at fixed tensile elongation of a polyurethane
strip in adhesive contact with itself, by contacting an upper strip 1 cm wide,
10 cm long and 3.175 mm thick together with a substrate 1 cm wide, 11 cm long
(i.e. u=1 cm) and 6.35 mm thick; hence, this joint was characterized by the parameter E_2h_2/E_1h_1=.5. The time required to impose the instantaneous tensile elongation was in any case inferior to 1 sec, a very short duration with respect to the
rupture time. Figure 5 shows the variations with time of the contact length ℓ for
different fixed elongations δ applied to the system, in the range 2-15 mm. Instan-

taneously, ℓ takes a value greater than its initial ones L and then a continuous decreasing is observed. Experiments were stopped after 10 min due to the limited film capacity of the camera.

Corresponding values of the crack speed calculated by eq. 10, with x deduced from eq. 9 and data of fig. 5, and the associated strain energy release rate G computed by eq. 4 are represented on fig. 6 on Log-Log coordinates. All the points fall on the same straight line with the slope .6, that corroborates the previous findings (ref.7) and consequently confirms eqs. 1 and 2 in which the function Φ for polyurethane samples varies as the .6th power of the crack speed, as expected. Taking into account $\alpha(\theta)= 4.75 \ 10^4$ SI units, previously assessed and corresponding to the room temperature $\theta= 23°C$ (ref.10), and data from fig. 6, the mean value of the thermodynamic work of adhesion that may be calculated by eq. 1 is $\overline{w} = 42 \ mJ/m^2 \ (\overset{+}{-} \ 2 \ mJ/m^2)$, a fairly low value probably due to the high humidity ratio (refs.18,19), so that w can be neglected in the left member of eq. 1.

Figure 7 shows variations, recorded by the tensile machine, of the force to which the experimental arrangement is subjected with time, in the same conditions as in fig. 5. As expected, as soon as a fixed tensile elongation is imposed, the corresponding force takes instantaneously a maximum value which is proportional to the initial uniaxial deformation and then continuously decreases. After complete delamination (x=L) the force remains constant and, generally, a readhering of the relaxed upper strip on the strained substrate is observed.

By numerical integration of the differential equation proposed (eq.1), taking into account eqs. 3 and 4 and the variation of the function Φ with the .6th power of the crack speed, it is possible to predict variations of the force during the shear delamination. The computed curves (heavy lines on fig.7) are in quite agreement with experimental results. It can be pointed out that the small vertical shift visible on some curves arises from a weak variation of w with respect to the mean value used for calculations. Moreover, the slight divergences observed for long times are probably due to dwell time effect resulting from the relaxation of stresses stored in the roughnesses at the interface (ref.16).

5 SPONTANEOUS DELAMINATION AT FIXED CROSSHEAD VELOCITY

The same experimental arrangement has been tested in a second set of solicitations in which the lower strip of polyurethane, used as elastic substrate, was submitted to various fixed crosshead velocities. In this case, the kinetics of separation is less easy to interpret as at fixed tensile elongation, because as in the tackiness problem (ref.10), it is due to the competition between increasing elongation δ at constant contact length ℓ (or constant relaxed peeled length

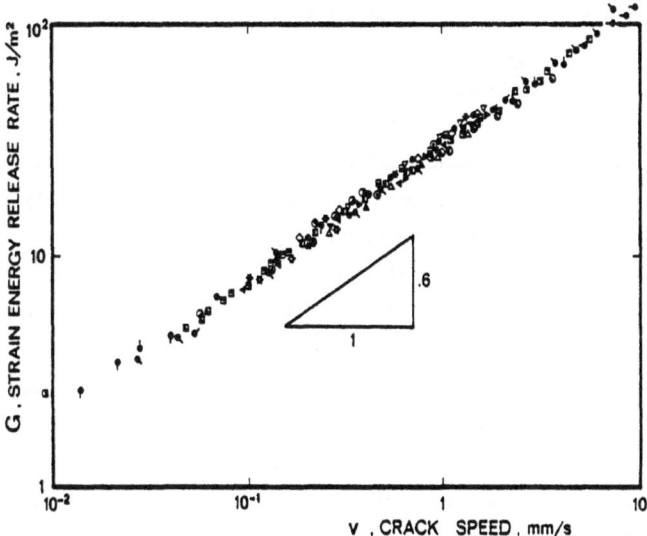

Fig. 6   Strain energy release rate versus crack speed for strips of poly-urethane in adhesive contact, for various fixed tensile elongations and crosshead velocities, lengths L and u, widths b and thicknesses h.

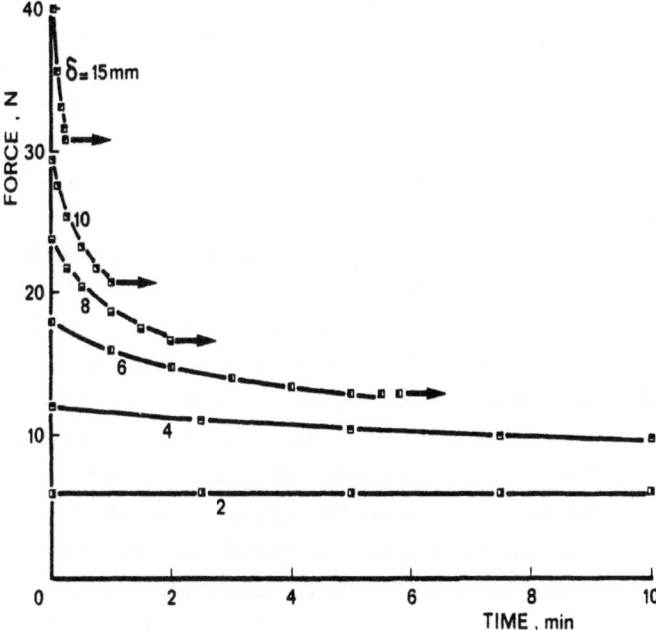

Fig. 7   Variations with time of the force applied to the system in the same conditions as in fig. 5. Experimental results and corresponding com-puted curves (heavy lines).

x) and decreasing ℓ (or increasing x) at constant δ for increasing G :

$$\frac{dG}{dt} = \left[\frac{\partial G}{\partial \delta}\right]_x \cdot \dot{\delta} + \left[\frac{\partial G}{\partial x}\right]_\delta \cdot \frac{dx}{dt} \qquad \qquad \ldots (12)$$

Variations of the measured contact length ℓ versus time are shown on fig. 8. At the beginning, as long as G remains smaller than w, the crack cannot start and the contact length ℓ increases with δ. Moreover, when G just becomes slightly greater than w, the crack begins to propagate with a very slow speed so that the contact length continuously increases for a short time and then decreases until the complete delamination is reached; obviously, the shortest rupture times correspond to the highest crosshead velocities.

Figure 9 shows variations of the force sustained by the system and recorded by the tensile machine, in the same conditions as in fig. 8. Due to the slower evolution of the speed $d\ell/dt$ just before the total rupture of the contact (fig.8), the force is seen to increase less rapidly. There after, as expected, the force linearly increases with time and corresponds to the continuous uniaxial deformation of the substrate.

Relations between the strain energy release rate and the associated crack speed, calculated by eqs. 4 and 11 respectively, are shown on fig. 6. They wholly confirm the variation of the function Φ with the .6th power of the crack propagation speed. As previously, it is possible, by numerical integration of eq. 1 and using eqs. 3 and 4, to predict the theoretical evolution of the force sustained by the adhesive joint. Computed curves (heavy lines on fig.9) are in good agreement with experimental results. Moreover, the curve A on fig. 10 shows that the time required to observe the complete delamination is a power function of the crosshead velocity.

In order to further verify the good validity of the model, other tests were realized at constant crosshead velocity ($\dot{\delta}$= 5 mm/min), as studies of the influence of the width b, of the initial contact length L and of the initial length u of the free end of the substrate. Experimental results and corresponding computed curves, calculated with w= 42 mJ/m^2, are compared on figs. 11, 12 and 13. It is observed that a change in contact width b, all other geometrical parameters remaining constant, does not alter the kinetics of delamination. Indeed, curves $\ell(t)$ are all superimposed, so that the complete rupture of the joint is observed after the same duration (fig.11). Moreover, as expected, the recorded force is, at every time, proportional to the width b. On the other hand, an increase in the initial contact length L delays the appearance of the total rupture of facing and the corresponding time is almost proportional to L, as shown on fig. 12, whereas the force recorded at the complete rupture seems to be independente of L.

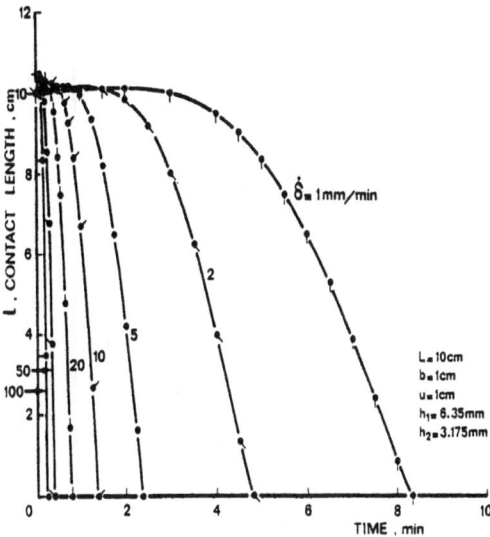

Fig. 8   Variations with time of the contact length between two strips of polyurethane for various fixed crosshead velocities imposed to the substrate, as in fig. 2B.

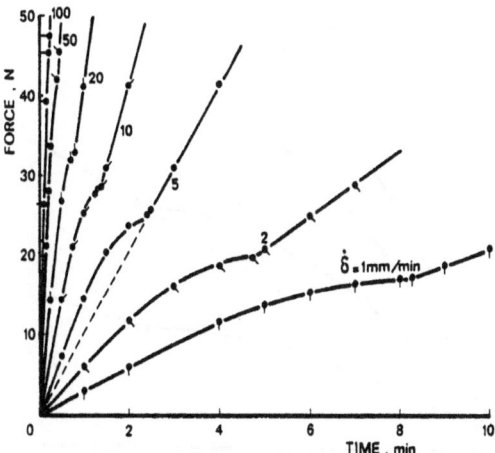

Fig. 9   Force versus time in the same conditions as in fig. 8. Experimental results and corresponding computed curves (heavy lines).

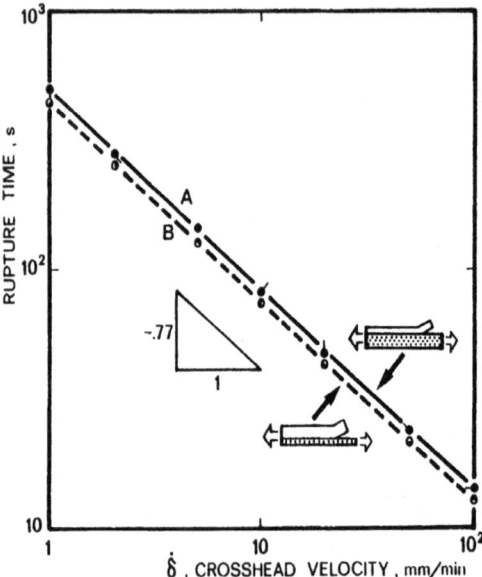

Fig. 10   Rupture time versus crosshead velocity for a lap shear joint made by contacting two strips of polyurethane. A: $E_2h_2/E_1h_1 = .5$; B: $E_2h_2/E_1h_1 = 2$.

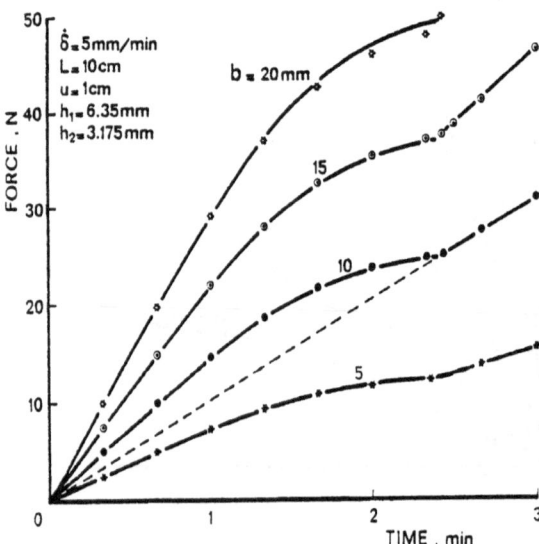

Fig. 11   Force versus time for various joint widths b. Experimental results and corresponding computed curves (heavy lines).

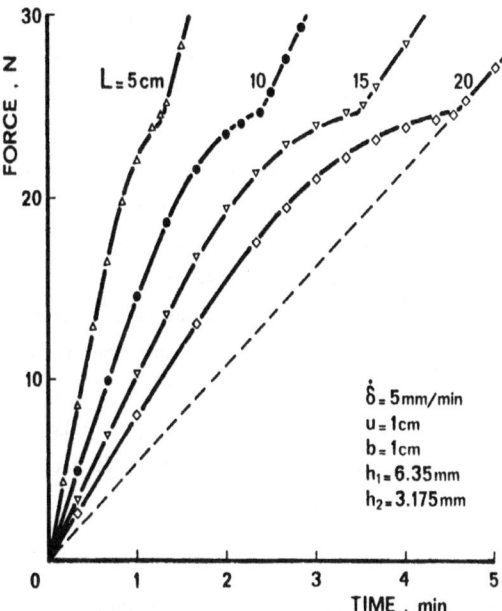

Fig. 12   Force versus time for various initial contact lengths. Experimental results and corresponding computed curves (heavy lines).

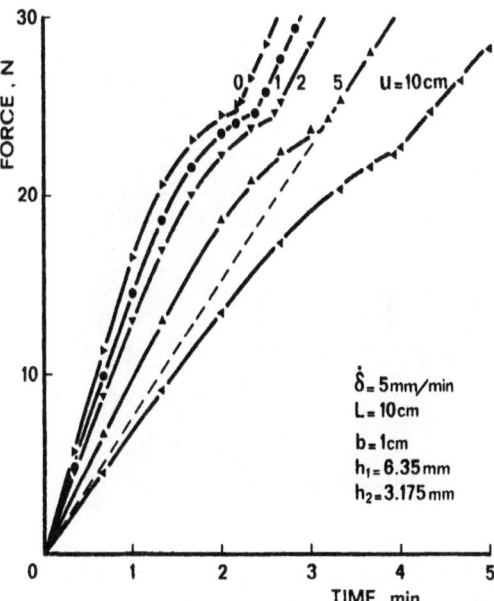

Fig. 13   Force versus time for various initial free end lengths of the substrate. Experimental results and computed curves (heavy lines).

Lastly, fig. 13 confirms that the initial length u of the free end of the elastic substrate plays a major part in the kinetics of propagation but only a minor ones on the force applied to the system.

During crack propagation the spontaneous peeling angle was measured. A slow increasing with time following by a sudden decreasing just before the total rupture was observed. Simple geometrical considerations associated with the incompressible character of the polyurethane ($\nu \approx .5$) enable one to relate the value of this angle to the uniaxial strain of the adhesive part of the upper strip (fig.2B). Indeed, to a first approximation, the flaking of the peeled strip may be easily ascribed to the variation of thickness between the two adhering and relaxed states and one can write : $\cos\psi = h'/h = 1 - \varepsilon/2$, hence $\psi \approx \sqrt{\varepsilon}$ if ε is not too high. Although the substrate is not a rigid body and is not perfectly plane in the vicinity of the crack tip, as shown on fig. 14, measurements of peeling angle confirm the prediction with an accuracy of better than 7%.

Fig. 14   View of the lap shear joint during crack propagation at fixed crosshead velocity.

Figure 15 shows, for comparison with fig. 9, variations with time of the force applied to the system at fixed crosshead velocity $\overset{\bullet}{\delta}$, if the adhesive joint is made by contacting a polyurethane strip 6.35 mm thick with a polyurethane substrate 3.175 mm thick, i.e. in the opposite situation of the previous lap shear joint studied, and in this case the ratio $E_2 h_2 / E_1 h_1$ takes the value 2. The main difference is the decrease of the force for a given $\overset{\bullet}{\delta}$ associated with its marked stabilisation before rupture of contact. This phenomenon seems to be characteristic of viscoelastic strips having a parameter Eh greater than the substrate ones (ref.23). Moreover, the time required to observe complete delamination is shorter as shown on fig. 10 (curve B). It can be pointed out that all tests realized at fixed crosshead velocity with polyurethane-polyurethane joints confirm, whatever the geometrical parameters (L,b,h_1,h_2,u), the variation of the rupture time with the -.77th power of the crosshead velocity. Consequently, this behaviour is a characteristic of the viscoelastic material tested, closely related to the variation of the function Φ with the .6th power of the crack propagation speed. Other experiments realized with adhesive joints made by contacting two strips of polyurethane with the same thickness h=3.175 mm lead to similar results and conclusions (ref.23).

A last set of experiments has been realized in testing composite lap joints made by facing polyurethane (E = 3.6 MPa, h = 3.175 mm) and natural rubber (E = 1.3 MPa, h = 2 mm) strips, every material being used as substrate or peeled strip (ref.24), so that the ratio $E_2 h_2 / E_1 h_1$ takes values 4.4 or .23. Hence, relations between the strain energy release rate and the length of the relaxed peeled strip in a test at fixed tensile elongation are quasi-similar to that given on fig. 3. Figure 16 shows variations of the contact length ℓ with time when the fixed crosshead velocity $\overset{\bullet}{\delta}$ = 5 mm/min is imposed to the system. For comparison, results concerning NR/NR and polyurethane/polyurethane joints are also given. As comparison between figs. 9 and 15 enabled one to predict, the rupture time is shorter if the peeled strip has a parameter Eh higher than the substrate ones (curves 1 and 2 on fig.16). Relations between the strain energy release rate and the associated crack speed, for the four joints studied are shown on fig. 17, on Log-Log coordinates. All points fall on straight lines, with a slope n'=.58 if the natural rubber is peeled and with a slope n=.6 for the polyurethane, as expected. Again, the power variation of the function Φ with the crack speed is confirmed, and obviously, the curve 2 on fig. 17 is similar to the one drawn on fig. 6.

Using eqs. 3 and 4 and results from fig. 17, the numerical integration of eq. 1 makes it possible to predict variations of the force applied to the system. Experimental results and corresponding computed curves are compared on fig. 18. So

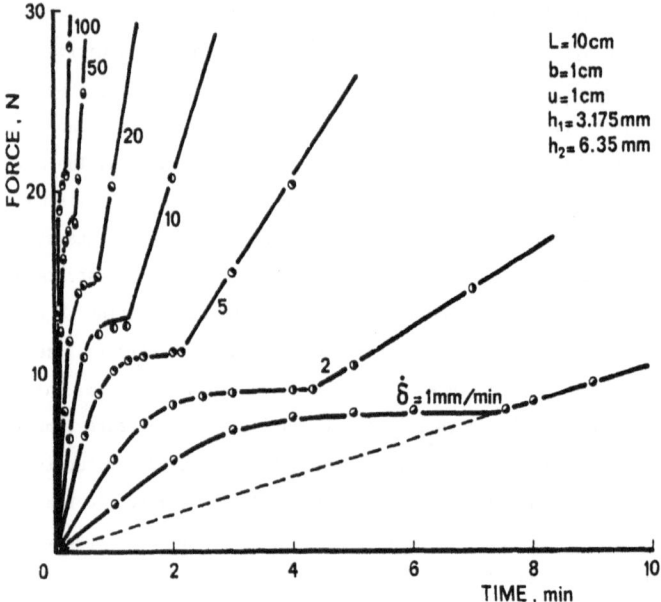

Fig. 15   Variations with time of the force applied to a lap shear joint
polyurethane-polyurethane with $E_2h_2/E_1h_1=2$, for various fixed crosshead
velocities. Experimental results and corresponding computed curves.

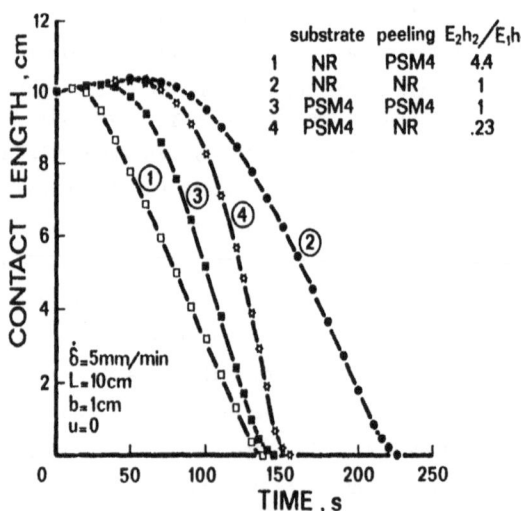

Fig. 16   Variations with time of the contact length of lap shear joints
made by contacting polyurethane and natural rubber strips, at fixed crosshead velocity.

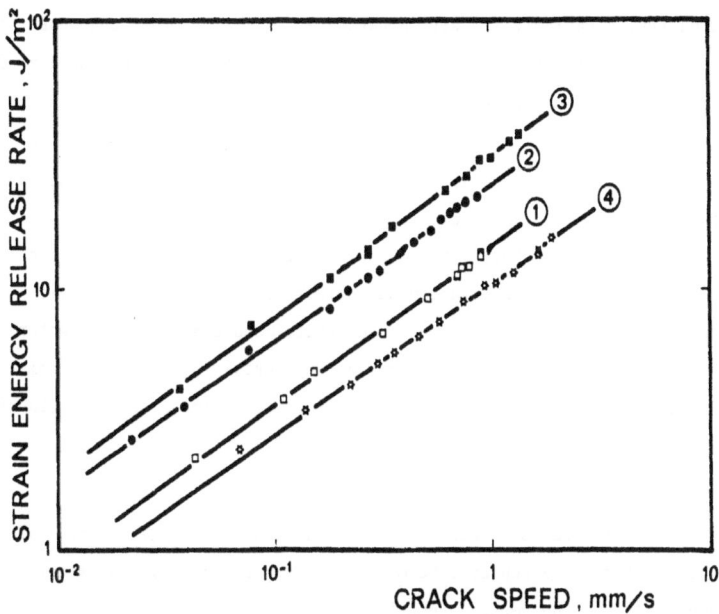

Fig. 17   Relations between strain energy release rate and crack speed corresponding to fig. 16.

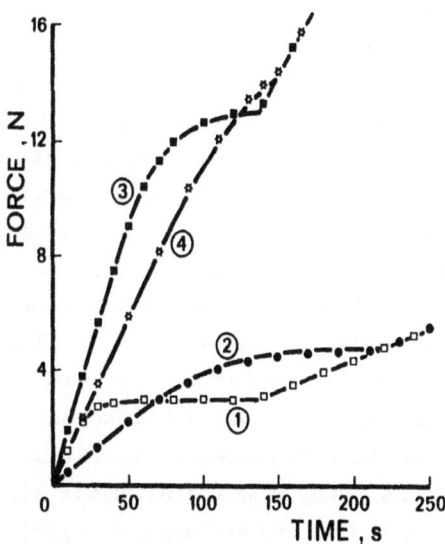

Fig. 18   Force versus time corresponding to fig. 16. Experimental results and computed curves (heavy lines).

it is confirmed that a peeled strip with a parameter Eh smaller than the substrate ones can sustain a higher strain before complete delamination. On the contrary, if the peeled strip presents a parameter Eh greater than the substrate ones, the force begins to increase and then remains constant (i.e. curve 1 on fig.18) until the complete delamination is reached, and the crack speed also as in a standard peeling test of a rubber-like material strip from a rigid substrate.

The whole of results proves the sound validity of the proposed model to entirely describe the kinetics of spontaneous delamination of an viscoelastic strip in adhesive contact with a deformable substrate, and due to the elementary nature of this model, it may be suitable for solving certain practical problems such as the ability of varnishes, paints and coatings to adhere on various elastic substrates.

6 CONCLUSIONS

The concepts of fracture mechanics may be used to study the kinetics of the spontaneous peeling of a rubber-like material strip in adhesive contact with an elastic substrate, by the general equation

$$G - w = w.\Phi(a_\theta.v) \qquad \qquad . \quad . \quad . \quad (1)$$

where G is the strain energy release rate, calculated from the elastic energy stored in the system, w is the thermodynamic work of adhesion and Φ is a dissipative function. In previous measurements of the kinetics of the adherence of rigid punches in adhesive contact with viscoelastic solids, it has been shown that the function Φ is a characteristic of the elastomeric body for the crack propagation in mode I and that it depends only on the temperature through the term a_θ (WLF shift factor) and on the speed v of the crack propagation at the interface. Experiments carried out with various lap shear joints made by contacting strips of polyurethane and of natural rubber also and submitted to instantaneous tensile elongations or fixed crosshead velocities, prove that eq. 1 remains valid in shear delamination geometry and that the function Φ is a power function of the crack propagation speed, as in normal approach measurements of the adherence.

Experimental determination of the function Φ thus makes it possible to predict the kinetics of propagation in all lap shear joints provided that the elongations remain purely elastic (with viscoelastic losses being left localized at the crack tip) and that the rupture is an adhesive rupture (with propagation of the crack at the interface). Equation 1 has been used to predict :the time required to observe the complete spontaneous delamination as a function of the fixed crosshead velocity, the size of the contact area and the force brought into play at every time during peeling. Particularly, it is demonstrated that a coating with a smaller parameter Eh than the substrate ones can sustain higher deformations and that in the opposite case, the force remains quasi-constant until the complete delamination is reached, if a crosshead velocity is imposed to the system. Experimental results verify theoretical predictions with an accuracy better than 4%.

ACKNOWLEDGEMENTS

The author would like to thank the Direction des Recherches, Essais et Techniques for the financial support given to this work (contact n° 83-1033).

REFERENCES

1. Kendall, K. 'Crack propagation in lap shear joints', J. Phys. D : Appl. Phys. vol. 8, 512-522, 1975.

2. Kendall, K. 'Shrinkage and peel strength of adhesive joints' J. Phys. D : Appl. Phys., vol. 6, 1782-1787, 1973.

3. Kendall, K. 'The effects of shrinkage on interfacial cracking in a bond laminate', J. Phys. D : Appl. Phys., vol. 8, 1722-1732, 1975.

4. Kendall, K. 'Control of cracks by interfaces in composites', Proc. R. Soc. Lond. A., vol. 341, 409-428, 1975.

5. Johnson, K.L., Kendall, K. and Roberts, A.D. 'Surface energy and the contact of elastic solids' Proc. R. Soc. Lond. A., vol. 324, 301-313, 1971.

6. Kendall, K. 'The adhesion and surface energy of elastic solids' J. Phys. D : Appl. Phys., vol. 4, 1186-1195, 1971.

7. Maugis, D. and Barquins, M. 'Fracture mechanics and the adherence of viscoelastic bodies' J. Phys. D : Appl. Phys., vol. 11, 1989-2023, 1978.

8. Barquins, M. 'Adhesive contact and kinetics of adherence between a rigid sphere and an elastomeric solid' Int. J. Adhesion and Adhesives, vol. 3, 71-84, 1983.

9. Barquins, M. 'Influence of the stiffness of testing machine on the adherence of elastomers' J. Appl. Polymer Sci., vol. 28, 2647-2657, 1983.

10. Barquins, M. and Maugis, D. 'Tackiness of elastomers' J. Adhesion, vol. 13, 53-65, 1981.

11. Maugis, D. and Barquins, M. 'Fracture mechanics and adherence of viscoelastic solids' in Adhesion and Adsorption of polymers, ed. L.H. Lee, vol. 12A, Plenum Publ. Corp., New York, 203-277, 1980.

12. Gent, A.N. and Schultz, J. 'Effect of wetting on the strength of adhesion of viscoelastic materials' J. Adhesion, vol. 3, 281-294, 1972.

13. Andrews, E.H. and Kinloch, A.J. 'Mechanics of adhesive failure' Proc. R. Soc. Lond. A., vol. 332, 385-399, 1973.

14. Williams, M.L., Landel, R.F. and Ferry, J.D. 'The temperature dependence of relaxation mechanisms in amorphous polymers and other glass-forming liquids' J. Amer. Chem. Soc., vol. 77, 3701-3707, 1955.

15. Maugis, D. and Barquins, M. 'Adhesive contact of sectionally smooth-ended punches on elastic half-spaces : theory and experiment' J. Phys. D : Appl. Phys., vol. 16, 1843-1874, 1983.

16. Barquins, M. 'Influence of dwell time on the adherence of elastomers' J. Adhesion, vol. 14, 63-82, 1982.

17. Greenwood, J.A. and Johnson, K.L. 'The mechanics of adhesion of viscoelastic solids' Phil. Mag. A., vol. 43, 697-711, 1981.

18. Barquins, M. 'Etude théorique et expérimentale de la cinétique de l'adhérence des élastomères' Thesis, University of Paris, 1980.

19. Roberts, A.D. 'Looking at rubber adhesion' Rubber Chem. Tech., vol. 32, 23-42, 1979.

20. Kendall, K. 'Transition between cohesive and interfacial failure in a laminate' Proc. R. Soc. Lond. A., vol. 344, 287-302, 1975.

21. Kendall, K. 'Thin-film peeling - the elastic term' J. Phys. D : Appl. Phys., vol. 8, 1449-1452, 1975.

22. Barquins, M. and Courtel, R. 'Rubber friction and the rheology of viscoelastic contact' Wear, vol. 32, 133-150, 1975.

23. Barquins, M. 'Kinetics of the spontaneous peeling of elastomers' J. Appl. Polymer Sci., 1984, in press.

24. Barquins, M. 'Sur le pelage spontané des élastomères' Comptes Rendus Acad. Sc. Paris, 1984, in press.